MODERN KNIFEMAKING for Beginners

Quarto.com

First Published in 2025 by New Shoe Press,
an imprint of The Quarto Group,
100 Cummings Center, Suite 265-D,
Beverly, MA 01915, USA.
T (978) 282-9590 F (978) 283-2742

Essential, In-Demand Topics,
Four-Color Design, Affordable Price
New Shoe Press publishes affordable, beautifully designed books covering evergreen, in-demand subjects. With a goal to inform and inspire readers' everyday hobbies, from cooking and gardening to wellness and health to art and crafts, New Shoe titles offer the ultimate library of purposeful, how-to guidance aimed at meeting the unique needs of each reader. Reimagined and redesigned from Quarto's best-selling backlist, New Shoe books provide practical knowledge and opportunities for all DIY enthusiasts to enrich and enjoy their lives.

Visit Quarto.com/New-Shoe-Press for a complete listing of the New Shoe Press books.

New Shoe Press titles are also available at discount for retail, wholesale, promotional, and bulk purchase. For details, contact the Special Sales Manager by email at specialsales@quarto.com or by mail at The Quarto Group, Attn: Special Sales Manager, 100 Cummings Center, Suite 265-D, Beverly, MA 01915, USA.

10 9 8 7 6 5 4 3 2 1

ISBN: 978-0-7603-9804-3
eISBN: 978-0-7603-9805-0

The content in this book was previously published in *A Modern Guide to Knifemaking* (Quarry Books 2018) by Laura Zerra.

Library of Congress Cataloging-in-Publication Data available

Page Layout: *tabula rasa* graphic design

Photography: Jessica Olivier unless otherwise noted. Page 6: courtesy of Discovery Channel, Page 33: courtesy of Mike Dion, Page 32: courtesy of Jeff Bailey, Pages 34 (lower), 50: courtesy of Paul Brach, Pages 51, 103 (bottom right): courtesy of Jeff Bailey, Page 103 (top right): courtesy of Travis Payne, Page 105 (top): courtesy of Mike Hawkridge, Page 105 (bottom): courtesy of Mike Jones, Page 89: courtesy of David DeAustin, Pro Tips photos not listed above: courtesy of the knifemaker

Illustration: Mattie Wells

Printed in China

Knifemaking and handling knives can be dangerous activities. Failure to follow safety procedures may result in serious injury or death. This book provides useful instruction, but we cannot anticipate all of your working conditions or the characteristics of your materials and tools. For your safety, you should use caution, care, and good judgment when following the procedures described in this book. Consider your own skill level and the instructions and safety precautions associated with the various tools and materials shown. The publisher cannot assume responsibility for any damage to property or injury to persons as a result of misuse of the information provided.

MODERN KNIFEMAKING for Beginners

Step-by-step instructions for forging your own knife from expert bladesmiths, including making your own handle, sheath, and sharpening

LAURA ZERRA

NEW SHOE PRESS

Contents

Introduction

Humankind has relied on knives for survival for thousands of years. The ability of our ancestors to make and utilize knives had a huge impact on our success as a species. According to anthropologists, primitive cutting tools were some of the earliest man-made objects. The first knives were fashioned from flakes of sharp rock systematically broken through the art of flintknapping to form a cutting edge. These sharp points were then affixed to various sizes and shapes of wood or bone to make spears, arrows, and utility knives. The primitive blades assured the survival and advancement of humanity by giving our ancestors the competitive edge they needed to make it in a harsh and unforgiving world.

In our modern world, most of us no longer have that direct connection to our own survival. Yet still, most people own and regularly use knives to complete such everyday tasks as cooking dinner. In a world where technology is constantly advancing and tools are replaced with gadgets, knives have remained surprisingly the same; nothing beats the efficacy and usefulness of a simple blade. The draw so many of us feel to knifemaking is tied to this vast history. It is part of our story. Our very success as a species has been intrinsically linked to the blades we use, and our relationship with the knives we carry reminds us of what it is that makes us human.

Over the years, I've learned what works for me in a blade and what doesn't. This search for the perfect blade to suit my needs inspired me to try my hand as a knifemaker. The ability to create the knife of my wildest dreams, and then change and improve upon it, is a constant and rewarding challenge. The techniques in this book are what worked for me in my learning process, but they are certainly not the only way to get the job done. As you spend more time practicing the art of knifemaking, you'll find what works for you and what modifications suit your budget and your style.

I encourage you to experiment with different types of metals and materials to find out what you like best and to seek out as many resources as you can to further your knowledge. While I began my knifemaking journey as a means to an end, I've found that the real beauty has been the process, allowing me to continue in the long succession of knifemakers who have made human history possible. Happy forging!

Survival in the wild with nothing but a knife taught me the importance of a good blade.

CHAPTER ONE

Tools of the Trade

Now that you've decided you want to be a knifemaker, it's time to set up a work space and make sure you have the right tools. As you first start out, don't be limited by the thought that you need to invest a ton of money to have the best tools and a huge work space. While investing in certain equipment down the line may make the process easier, most of the tools you need to start making knives are probably in your garage right now. In this chapter, we'll go over what kind of space you need to set up your workshop and what tools you need to complete the different tasks.

Setting Up a Shop

Someday, in a perfect world, I'll have a huge and immaculate work space set up with every tool I could ever dream of. My reality at the moment is quite the opposite. As a self-proclaimed nomad who rarely spends more than a few months in one place, I find myself making do with what I have available—whether it be using someone else's shop or setting up a work space in whatever place I can find at the time. There is no excuse for not being able to start on your journey into knifemaking, regardless of your current availability of space. Where there's a will, there's a way; and with a bit of ingenuity, you'll be grinding metal in no time.

If you have a work space already set up in your garage, great. With a few modifications, you can get everything you need set up there to start making knives. If you don't, there are plenty of other options. From sheds to lean-tos to tarped-in shelters, I've effectively used all sorts of different locations as a makeshift shop.

The first requirement for a work space is having the ability to get it dirty. As you're systematically taking metal away from the piece of steel that will become your knife, the metal dust that you're removing will find its way onto and into everything. Add in using different glues, a forge, and working wood and leather, and keeping the place clean will become nothing short of impossible.

Additionally, your space should have ventilation. While wearing a respirator will help, you won't want to have to wear it all the time. Being able to air out the smoke from working metal in the forge and the fumes from glue is important.

You'll also need access to electricity. I've used a generator in a pinch. While people made knives for thousands of years with absolutely no electricity, power tools make this process infinitely easier.

Working in my dream shop, although this one happens to belong to New Hampshire–based blacksmith Steve Ash

My current shop is set up in an old sugar shack in Maine. In the spring, it is used for boiling down sap to make maple syrup.

Most importantly, pick a location that's protected from the elements and is easily accessible. Making a knife is a process, and being able to put down your tools and come back to it without much effort makes it easier to get in shop time. You can have the best shop in the world, but it will do you no good if you never spend time in it.

Safety Equipment

Every time I think about safety, my mind conjures up images from cheesy old safety videos. There's an old saying from one made in the 1940s that I quite enjoy: "Don't learn safety by accident." As a survivalist, I find I encounter a fair amount of danger in my life. This in no way means I'm careless. I like to save the calculated risk for adventures and protect myself from unnecessary hazards in the shop.

Safety Glasses

With the use of most power tools, there is the risk of an unwanted object flying into your eye. This is especially true when grinding metal. Metal shards in the eye are unforgiving, painful, and gross. I'll never forget how horrified I was as a kid trying to help my dad get a piece of metal out of his eyeball with a magnet after an accident in the garage. I've been wearing glasses or a face shield in the shop ever since. It's like they always say, "To learn about eye protection, ask someone who has one."

Hearing Protection

The very first knifemakers lived in a much quieter world. Loud sounds can be hard to get away from in modern life, and the effects can be serious. Prolonged exposure to any elevated sound actually kills the nerve endings in the inner ear and causes permanent hearing loss. In my opinion, your time in the shop should enrich your life instead of adding unnecessarily to the slow demise of any part of your body. I use hearing protection whenever I use power tools, or if I am going to be spending a long time hammering. I've always said, "Hearing protection is a sound investment."

Respirator

When grinding metal, the tiny metal flakes you remove become an airborne dust. Inhaling this dust over time can cause a disease called siderosis, or "welder's lung," which can have very serious complications. The annoyance of wearing a respirator is definitely worth it for your long-term health. After all, it's like they always say: "Working safely is like breathing. If you don't, you die."

Dress for the Occasion

Working around a forge can be incredibly hot. The same heat that is welcome in the winter can be stifling in the summer. I've been guilty of wearing shorts in the shop on hot days and usually come away from it with a few burn marks from hot sparks flying from my hammer and hitting my legs. If you're wearing shorts, it's possible for these sparks to land right inside of your boot and give you a good burn. When I was working as a farrier, I learned the hard way one of the reasons some knifemakers choose to wear leather bibs. As I was using a belt sander and grinding a horseshoe, the hot metal dust collected in a wrinkle in my sweatshirt. As the amount of dust and the temperature increased, it actually lit my shirt on fire.

Some blacksmiths also choose to wear gloves to further protect their hands. I've found that I can't grip the tools quite as well when I'm wearing gloves, but it can be a bit uncomfortable when you're working so close to a heat source. Your choice in attire really comes down to what amount of discomfort you're willing to deal with. You've been warned; proceed at your own risk.

Tools

Knifemaking involves more than just the skill of metalworking. As you progress in the process of making your knife, you'll be implementing other skill sets as you use different materials to add a handle and make a sheath. The following tools are a basic starting point for making a fixed-blade bushcraft-style knife using the process outlined in this book. Depending on your experience and familiarity with different tools, I encourage you to experiment and improvise to find what works best for you.

A solid workbench is a must-have in any knifemaking shop. Your bench should be a comfortable height to work at without having to bend over.

Workbench

A workbench will not only be your primary work space but will also be used as a mounting platform for other tools, such as your vice. You can improvise with a sturdy table or make a simple platform out of scrap wood. It's very important that your workbench be stable so that it doesn't move around when you're working on your blade. If anchoring it to the floor or wall isn't possible, try to add weight on the bottom to create a solid base with as little movement as possible.

While the angle grinder can be a great time-saver, make sure you use it with caution. A spinning wheel that can cut through metal will make easy work of your leg if you let it.

Angle Grinder

While not absolutely necessary, this tool is multifunctional and can be a great time-saver. I use an angle grinder any time I have to cut steel, and it can also be useful in profiling and grinding your blade. You'll need a variety of different discs to perform different tasks, including metal cutting and grinding discs.

Drill

I prefer to use a drill press for all my hole-making needs when I have one available to me. I find I can keep my holes straighter than when using a hand drill, but either one will work fine.

Files

I use files designed for both metal and wood at various stages of making a knife. While I do the bulk of material removal with power tools, files are very convenient for small tasks such as quickly removing a metal burr or doing some fine-tuning on the handle. Having a variety of files around gives you the option to slow things down and really take your time, especially when working on fine details.

Belt Grinder

If you're going to invest in one tool as a bladesmith, I would recommend getting a good belt grinder. While you can accomplish many of the same tasks by using an angle grinder, I find the belt grinder easier to use and learn on. From grinding bevels to shaping the handle to finishing, it has a multitude of purposes and is the tool I use the most in the shop. While it's not necessary, it is a huge time-saver and will earn its keep if you plan on regularly making knives.

You can pick up a brand-new 1 × 30–inch (25 × 762 mm) belt grinder relatively cheaply. While not as rugged as its larger counterparts favored by professional knifemakers, it is affordable and functional for many things. The next step up is the 2 × 72–inch (51 × 1829 mm) belt grinder, which is more of a monetary investment but is also hardier than a small grinder. I use a variety of different belts and recommend getting several different metal grinding belts in grit 60 through 600.

The biggest advantage I find in using in a belt grinder over the angle grinder is the ease and control I have moving the blade over something that is stationary. When you use an angle grinder, you constantly have to readjust the clamping of your blade. I find it easy to lose track of what you're doing and of the overall picture.

Position the grinder at about navel height in your shop so you can look down over the belt and see what you're doing.

Quench

A quench is a container for holding liquid to cool down your metal. Your shop should have two quenches, one containing water and one containing vegetable oil. When quenching a blade in oil after exposing it to high heat, there is the possibility of it "flaring up," or having the oil start flaming. Make sure your quench is fireproof and has a lid so you can cut off the flow of oxygen if you need to. An old metal coffee can is a great improvised quench, provided that your whole blade will fit in it.

Forging vs. Stock Removal

There are two main schools of thought in knifemaking: forging and stock removal. Forging involves employing the techniques of blacksmithing and heating metal in a forge to rearrange the steel into a blade. Stock removal skips the blacksmithing and heads right to the grinder, shaping the blade by systematically removing steel. These two processes aren't necessarily used exclusively; many blacksmiths will finish up a blade on the grinder, and people who rely mostly on stock removal will occasionally turn to the forge for certain tasks. Each method has advantages and disadvantages.

There's a very traditional feel to blacksmithing that immediately comes to mind when most people think of knifemaking, but it also demands a greater investment of time and space to learn the process and set up the shop. As you become more proficient, it's possible to spend five minutes on the anvil and save thirty minutes of grinding. Before investing in a fully set-up smithy, look for a local blacksmith in your area so you can get your feet wet and see if it's for you. Many smiths are more than happy for you to come in and observe. Don't be surprised if you end up hammering on some metal before the end of the day.

Like many things in knifemaking, it ultimately comes down to personal preference. While there's plenty of discussion as to which process is superior, I've seen beautiful, functional knives created using both methods. Setting up a forge regardless of what method you favor can be advantageous for heat-treating, but it's possible to achieve the same results in a pit fire (outlined in chapter 6). If you don't want to set up a forge right away, skip chapter 4 on blacksmithing, cut out your blade profile with the angle grinder, and head right to grinding in chapter 5.

Many knifemakers incorporate both blacksmithing and stock removal methods to make their blades.

Blacksmithing Tools

The first blacksmithing setup I ever used belonged to my friend Slim Sharp. As his name implies, he has some height on me. At 6 feet 3 inches (110 cm), he is a full foot taller than I am. Starting out, I had no idea how much our height difference was affecting my experience and abilities as a blacksmith. I found myself tiring quickly and feeling sore, and I thought I needed to grit my teeth and muscle through the process. While blacksmithing does certainly involve putting in some elbow grease and mastering technique, a lot of the trouble I was having was because I wasn't using my body mechanics to my advantage. Working in my own shop, everything became easier. By paying attention to the details and setting up your shop to your own personal specifications, you can be much more efficient. While you can still use someone else's smithy, there is no one-size-fits-all setup; everything should be set up in accordance to your own body measurements.

The sheer number of different tools in a professional blacksmith shop and the unique and incredibly specific purpose each one serves can be overwhelming. As you start out, you don't need to have everything. In fact, most blacksmiths will tell you that an important part of your journey is to learn to make each and every one of these tools along the way. Even if you don't want to become a master smith, the time you can save by using the forge to start your blades is enough to make many knifemakers dabble in the art. The following tools are the bare minimum to get you started.

Hammer

The hammer is an iconic representation of the blacksmith. It serves to form the connection between you and your steel, and it should feel like an extension of your own body. Even if I am headed to someone else's shop, I always try to bring my own hammer. I like to imagine what TSA thinks as they scan my bag and wonder what I'm doing traveling with clothes, toiletries, and a hammer.

The head of your hammer should be a comfortable weight somewhere between one-and-one-half and three pounds. It should be light enough to use without causing muscle fatigue but heavy enough to move metal. Find the balance that works for you. The handle on your hammer should be roughly the same length as the distance between the tips of your fingers and your elbow. Your body is used to working with that length, and it helps you become more accurate in less time. Shave the handle down to fit your hand snugly. You should be able to grasp it easily without having to use a death grip. Hammering is all about being easy and efficient, so take the time to make your hammer as comfortable as possible.

Tongs

In one hand you'll have your hammer, and in the other hand you'll have your tongs. As the much less famous sister of the pair, the tongs are nonetheless important. You want to be able to hold on to the steel you're working comfortably and securely. While pliers will do in a pinch, a good set of tongs will make your life easier. The most important thing in selecting a pair of tongs is that the jaws can firmly grasp the thickness of your steel. As someone with very small hands, I also like to find a pair I can grasp easily without straining my fingertips too much.

Setting up your anvil height to your own body mechanics assures you get the most power out of every hammer swing. By having the anvil meet the hammer where your knuckles naturally fall, you aren't reaching to hit your steel or stopping short before you get in your full swing.

Anvil

A good anvil is pretty expensive to buy brand new, but with some searching you can usually find them for sale secondhand or even sitting in the corner of an old barn gathering cobwebs. Anvil technology hasn't changed a whole lot over the years, and I prefer to use an anvil that has some stories and imperfections. The main thing to watch out for is any major indentations or chips that will cause problems on the working surface.

While many master blacksmiths prefer to use an anvil over one hundred pounds, a knifemaker can get away with something much smaller. Ideally, I still like using an anvil that weighs at least fifty pounds. The lighter the anvil, the less efficient you will be because more of the energy from your hammer strike will go into the anvil instead of the piece of steel you're trying to work. The bigger the anvil, the easier things become.

An anvil is specifically shaped for many of the processes involved in blacksmithing and includes such features as a horn and a hardy hole for inserting specific attachments. If you just want to get into the basics of pounding metal, any large chunk of steel can be used to improvise. I've heard of people using pieces of railroad tracks, broken forklift tines, or machine shafts turned on end and secured to a heavy base. People have practiced the art of blacksmithing without access to perfect store-bought anvils for thousands of years. Find what works for you.

Anvil Stand

Your anvil needs to be firmly anchored to a solid base, or anvil stand. Anvil stands are made from a large sturdy piece of either metal or wood. I prefer to use wood, as it is generally easy and cheap to obtain. Hardwood tends to work better, as it's less prone to splitting. Oak, hickory, maple, and elm are all good choices. Make sure you cut the ends parallel to each other to make sure that your anvil sits flat on the stand. If possible, secure your stand to the ground to prevent wobbling.

To determine how tall your anvil stand should be, stand with your arm at your side making a fist. Your anvil should end up sitting just below where your knuckles fall. If you measure the distance between your knuckles and the floor, then subtract the height of your anvil, you should have a measurement for the height of your anvil stand.

As you are setting up your anvil, make sure to set it up within a few feet of your forge. You'll want to be able to have enough room to access the forge but be able to get to the anvil quickly to prevent unnecessary cooling of the metal.

Hardy Cut

There are several tools and attachments that will be useful as you continue your journey into blacksmithing. The one I use the most when knifemaking is the hardy cut. A hardy is a kind of tool with a square shank that fits into the hardy hole on top of the anvil. The hardy cut can be used to quickly cut a piece of metal off the piece you're working without having to cool your metal and use other tools. I use the hardy cut when I want to cut in the tip of a knife instead of pounding it in, which takes much more time and skill. You can also cut off any extra handle length that you find as you work. Most blacksmiths make their own hardy cuts, but you can find them secondhand quite easily. While it's not essential to knifemaking, it is an incredibly useful tool. I recommend picking one up if you come across it.

Forge

There are two main kinds of forges: propane-fired and coal-fired. They both have advantages and disadvantages. Either way, they'll both heat your metal to working temperature and get the job done.

Coal Forge

There's something to be said about the history behind a coal-fired forge and the ancient feel of pulling glowing metal from a pile of smoking coals. They're quieter than a propane forge and extremely versatile, and they make it easy to get localized heat. They can also be a bit challenging for a beginner. There is an art to running and using a coal fire. They require more maintenance, and it's more likely that you accidentally overheat and burn your work. They also take up more space and are less portable. Perhaps the biggest issue with coal-fired forges is that, due to the nature of coal, they are pretty dirty to run. The coal itself gets black dust everywhere, and the smoke can annoy any neighbors that happen to be downwind.

Propane Forge

Propane forges are clean burning, require less training and start time to use, and don't need the constant maintenance of the coal forge. They don't have the same old-school feeling that a coal forge does, but they're convenient and portable and won't make enemies out of your neighbors. That being said, they're also noisy and a bit ornery and require proper ventilation to avoid the risk of carbon monoxide poisoning. Unlike a coal forge, you're limited in the size of the project you undertake, as you can only work a piece of metal that fits within the dimensions of the chamber.

So what kind of forge should you set up in your shop? When you're first getting started, I recommend using the two-brick propane forge. It's cheap and easy to make and use, and it doesn't take up much of your work space. While it has a very small chamber, it's fully functional for making knives and is a great forge to learn on.

As you get into blacksmithing, you may find that you want a forge that is more durable and can accommodate bigger projects. Ultimately, the forge you'll use in the long run will be dependent on the space you have and your own personal preference. Spend time using both kinds of forges so that you can get a feel for what you like best.

A simple hand crank coal forge

> If you like using the process of blacksmithing in your knifemaking process, you may want to think about building a bigger forge for long-term use. I've seen creative yet fully functional forges built from materials found at the local dump. Once you understand the concept of a forge, you can keep your eye out for scrap material you can use to build a unique forge of your own.

An example of a propane forge

Building a Two-Brick Forge

While building your two-brick forge, you'll be creating a fair amount of dust from working your firebricks. Make sure to complete the project in a space with good ventilation that you don't mind getting dirty. The firebrick dust can damage your lungs, so wear your respirator.

MATERIALS AND EQUIPMENT

2 soft firebricks (9″ × 4½″ × 2½″ [22.9 × 11.4 × 6.4 cm])

metal cloth (12″ × 24″ × ¼″ [30.5 × 61 × 0.6 cm])

metal cloth wire

furnace cement

16.9 oz (500 ml) plastic water bottle

ruler

marker

half-round file

wire brush

wire cutters

drill

bit

respirator

propane hand torch

1. Mark out the chamber.

Stack your two bricks on top of each other and flip them on their ends. Using the ruler, find the center point of the two bricks and mark this spot. Place the water bottle centered on this mark and trace the bottom of the bottle. Repeat this process on the other end of the bricks. This cylinder will be the chamber of your forge.

2. Carve the chamber.

Unstack your bricks and lay them on your work space. You'll be removing a half-circle of material from the inside of each brick. Use your file to carve out the chamber, one side at a time. Be careful handling your firebricks, as they are very soft and break easily. Use your wire brush to clean out your file as it gets gummed up with the dust.

3. Drill the torch hole.

Once you have the two firebricks carved out, place them together and make sure they make a hollow cylinder in the middle. Make any touch-ups you need to. It's important to have a circular chamber so that the flame can swirl in your forge. Using your drill bit, drill a hole in the top brick, toward the back of the brick and angled slightly toward the front opening.

4. Reinforce with cement.

Spread the furnace cement in between the two bricks where they meet. Also, coat the inside with a thin layer. Let the cement dry for a couple of days.

Mark out your bricks to designate the area you'll be carving out. You can make the chamber slightly larger if you need to, but don't make it too big, as your forge won't heat up as well.

5. Wrap with metal cloth.

Take your metal cloth and fold it around your bricks. Cut off any excess and wrap the edges around the ends. Make note of where the hole in the side of your forge is. Using your wire cutters, cut out a piece of the metal cloth so that you can access this hole. Use your metal cloth wire to secure the cloth around the bricks. The soft firebricks have a tendency to crack, and this will help hold your forge together.

(continued)

Drilling the torch hole. Test the tip of your torch to make sure it fits.

The forge heating up. The flame should swirl around inside the chamber, as this will cause the heat to wrap around your work and give the most economical distribution of heat

6. Attach torch.

Set your forge on a non-flammable surface, as it will get very hot. Insert the tip of the propane torch into the hole in the back of your forge. You'll have to find a way to arrange your forge so that your torch can be vertical and the tip fits easily into the hole. You can use a few extra firebricks to prop up your torch and elevate your forge, or improvise with C-clamps or anything that is non-flammable and will keep your setup steady.

7. Fire up the forge.

Turn on your torch and check out your new forge. It's normal to get a few cracks on the first firing, but the metal cloth should hold everything together. You may need to adjust the angle of your torch slightly to get the right heat, so test it out on a piece of scrap metal before you try it out on your blade.

PRO TIPS: ALAN FOLTS

LOCATION: PALM BAY, FLORIDA

Alan Folts is a full-time custom knifemaker and a multitalented artist. Though he dabbled in painting, drawing, and multimedia, it was his skill as a sculptor that drew him into the world of knifemaking, where he found his calling. In addition to designing and making his own knives, he has made historical reproductions of blades. Alan enjoys teaching and mentoring people new to the knifemaking world and has collaborated with production knife manufacturers on licensed productions.

Knife style: "Modern Everyday Carry (EDC), with the use of a variety of materials to give the design flair."

How he got his start: "The first time I worked with metal was when I was about eight years old on my grandfather's farm. I would take apart old tools and reassemble them. I always had a knack for metal. I learned to mold metal and started working with different metals in my father's garage. I started building knives twenty-six years ago. My previous experience working in copperworks, metal casting, jewelry, and armor gave me plenty of experience to get me started on knives."

Best tip for a beginner: "Take your time and build things with control in mind. Use known materials so you can repeat and control your processes, and learn your process before you worry about selling any of your knives for profit. Build what you love, and love what you sell!"

On setting up a shop: "ORGANIZE! As you set up your shop, take the time to make your space ideal, safe, and comfortable. A good environment will make the work you do more enjoyable."

A knife made by Alan Folts

CHAPTER TWO

Designing a Knife

knife \ˈnīf\ n: a cutting instrument consisting of a sharp blade fastened to a handle (Merriam-Webster)

At its most basic definition, a knife is simply a cutting tool attached to a handle. And yet, there are thousands of different knives available on the market. How do you ever decide what the best kind of knife is? Knives are a very personal thing. We all have different tasks we need our blade to accomplish.

The best advice I've ever received about finding your own perfect knife is to use as many different knives as possible. When you begin to recognize the qualities of a blade that work for you, you can seek out other similar knives and further refine your exact specifications. By better understanding the different features and variations of a knife, you can quickly begin to recognize which specific features work for you—and which ones don't.

Designing a blade in the shop by the light of an oil lamp. It can be useful to design your knife in a spot that helps set the mood for creativity.

Knife Anatomy

It's good to know some basic terminology as you start to delve into the knifemaking world. I've always found that the more specific information I know about something, the more I start to pay attention to details I've never noticed before; I start to see it with new eyes. Learning basic knife anatomy will help see through the eyes of a knifemaker and assist you as you start to design your own blade.

This book focuses on fixed-blade knives as opposed to folding knives. A fixed blade consists of a cutting edge with a handle that doesn't fold and uses a sheath to protect the edge. A folding knife, or "folder," uses a pivot to fold the knife back into the handle.

Cutting edge, also known as the sharp edge or blade belly: The working part of the knife, which is ground to a sharp edge

Tip or point: The sharp point at the foremost part of the blade

Heel: The part of the blade that is closest to the bolster

Spine: The thickest part of the blade, it runs along the back of the blade and provides for its strength

Handle: The part of the knife that is held

Scales: The part of the handle that attaches to the tang to form the blade's handle

Grind (or bevels): The area of the knife that has been ground down to create the edge

Tang: The unsharpened extension of the blade to which the handle is attached

Butt: The very end of the handle

Handle fasteners: Often screws or rivets used to attach the handle to the tang

Guthooks are a popular feature on many hunting knives. As an avid hunter and butcher, I find them more gimmicky than useful. While they look tough, it is incredibly hard to keep the guthook sharp enough to be efficient at its job. The same result is easily achieved by using a sharp knife and a steady hand. I find the guthook more of a hindrance than a help, catching and snagging on things when I don't need it.

Guard: A general term for the piece between the handle and the cutting edge designed to prevent the fingers from slipping forward and reaching the cutting edge

Quillion: Another term for the guard on a knife, although it can be used to specifically refer to the pieces projecting out from the handle that actually stop your fingers.

Pommel: The specific piece that caps the end of the butt or attaches the tang. This part of the knife is sometimes used for hammering or striking. Often used interchangeably with butt.

Bolster: The thick "bolster" or band between the handle and the cutting edge. This creates a transition from the handle to the blade and helps to prevent your hand from being cut on the cutting edge.

Ricasso: The unsharpened length of blade between the handle and the start of the cutting edge

Choil: The small cutout on the base of the edge of the blade closest to the handle

Lanyard hole: A small hole, usually in the butt of the knife, that can be used to attach a cord

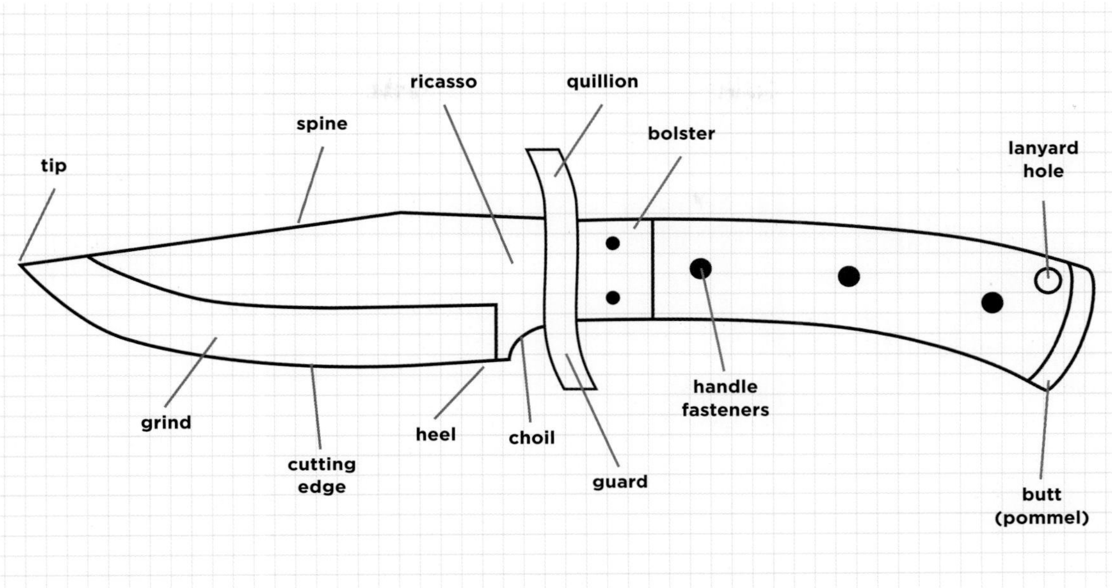
ricasso
quillion
spine
bolster
tip
lanyard
hole
handle
fasteners
grind
heel
choil
cutting
edge
guard
butt
(pommel)

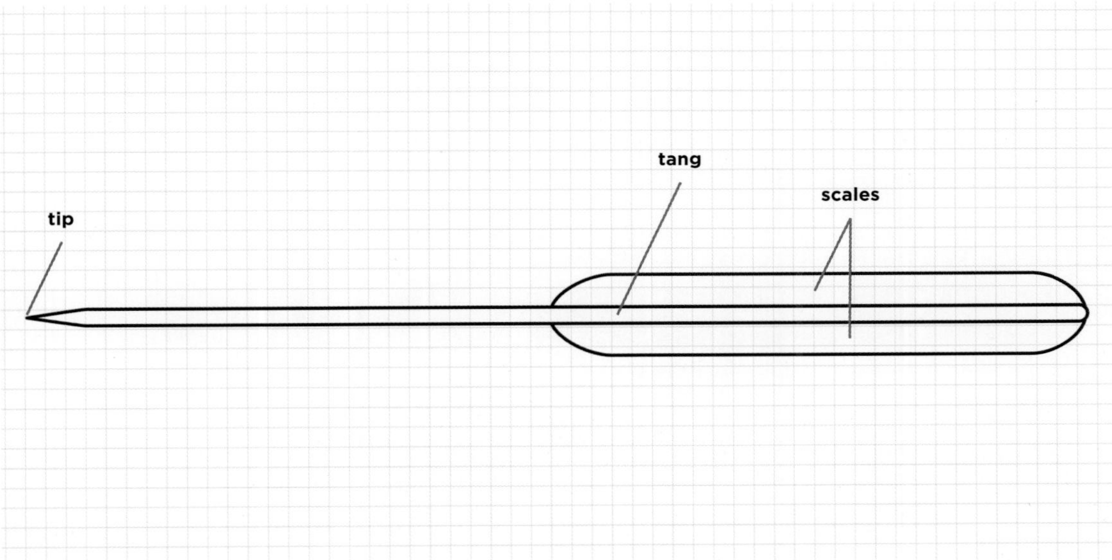
tang
scales
tip

Tang

Fixed blades have a tang, which is an extension of the blade that connects it to the handle. The tang helps give the knife its strength, balance, and weight. There are a few different kinds of tangs, depending on what you're trying to achieve in your blade.

When making a bushcraft knife, a full tang is preferable due to the durability and strength it gives to the knife. A full tang extends the entire length and diameter of the handle. The handle is made by attaching scales, or pieces of material, to either side of the tang, often using a combination of pins and glue.

There are several different kinds of partial tangs. These are all considered weaker than a full tang but serve to lighten the knife significantly by reducing the amount of steel in the finished blade. A push tang is made by having the steel taper off from the thickness of the blade and then pushing this smaller end into the handle material and securing it with glue. I often see hunting knives with antler handles made this way. Though I absolutely love antlers and they make for a beautiful knife, I wouldn't trust this style of tang on a knife I was using in a survival situation. The glue or adhesive has a tendency to let go when you most need it, especially if they age or are exposed often to the elements.

The project knife in hand. This blade has a drop point profile and a full tang.

Hidden tangs are another variation of a partial tang. They often run the full length of the handle or even run slightly longer so they can be screwed into or otherwise attached to a pommel at the butt of the handle. By using this kind of tang, you can avoid the look of pins on the handle and create a different aesthetic.

Skeletonized tangs are often utilized when trying to make a lightweight knife. Pieces of the steel in the handle are cut out to lighten the blade while leaving the outline of the full handle. Often, these are left bare or covered in parachute cord to create a makeshift handle that can also be removed and used for many purposes in a survival situation. This is a common style favored by "prepper" survivalists who want to be ready for the worst-case scenario at any given moment.

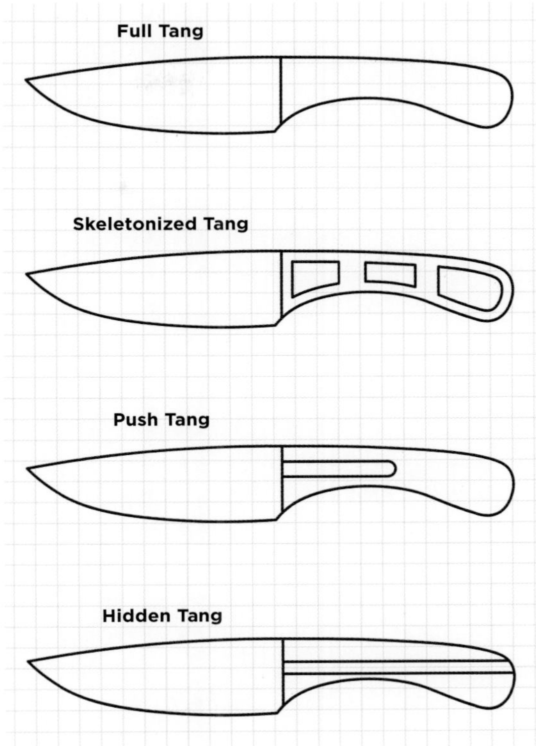

The different styles of knife tangs

Blade Profiles

Different knives will be better suited for different tasks, depending on their blade profile. The profile is the term used to describe the overall shape of the blade and gives the blade much of its look. Learning the basic kinds of blade profiles can serve as a guide as you're designing your blade based on the specific functions you want it to excel at.

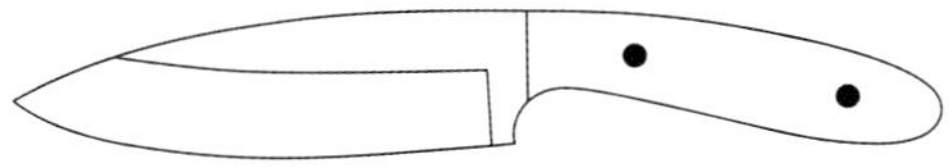

A design for a knife with a drop point profile

One of the most common blade profiles is the drop point, which is favored as the best style for survival and hunting blades. A drop point is characterized by the spine of the blade "dropping" down from the handle to the tip, with the tip ending up at the center axis of the blade. The spine extends the full length of the tip, which makes it much stronger and less prone to breaking. It has a large belly, which makes for a great skinning knife; and the positioning of the point makes it less likely you'll accidentally make any unwanted punctures in an animal. It also makes a great knife for carving.

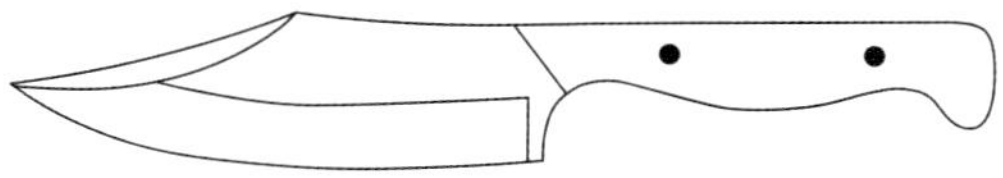

An example of a clip point blade design

A clip point blade is another common type of blade profile and is named after the "clipped off" appearance of the tip of the blade. The tip is sharper and thinner and, therefore, more suited for stabbing and piercing. However, this thin tip also makes it much weaker than a drop point blade, and it has a tendency to break more easily.

A tanto profile is sometimes used in the design of military-style, fighting utility knives. This profile has a spine that slopes slightly down to a point that is sharp and angular. This makes for a blade with a tip that is great for piercing, stabbing, and fighting, as well as general utility.

A sheepsfoot profile is characterized by a sharply sloping, blunt tip that meets at the blade's long cutting edge. While not a good all-purpose blade profile, this design has its purpose when you want to avoid accidentally stabbing an object while cutting. This is the style of blade frequently carried by first responders who have to perform tasks such as cutting off a seat belt. Sailors often carry these blades as well to avoid cutting themselves in rough seas as well as to avoid accidentally puncturing a sail.

A spearpoint design has a point that meets in the middle of two symmetrical sides, much like the tip of a spear. A spearpoint can have either one or both edges sharpened and has a very strong and sharp point. This feature makes it a great knife for piercing or stabbing. It's very difficult to do any fine carving or detail work with, and it isn't a practical everyday-use knife. People seem to think they could easily strap the blade to a stick and use it as a hunting spear in a survival situation, but that is easier said than done and is more likely to result in a broken or lost knife than dinner. In my opinion, this is a good blade design for self-defense but isn't the best choice for an overall outdoor knife. Take a blade more suited for the job of survival on your backcountry trips and stick to bone or rock points for your spears.

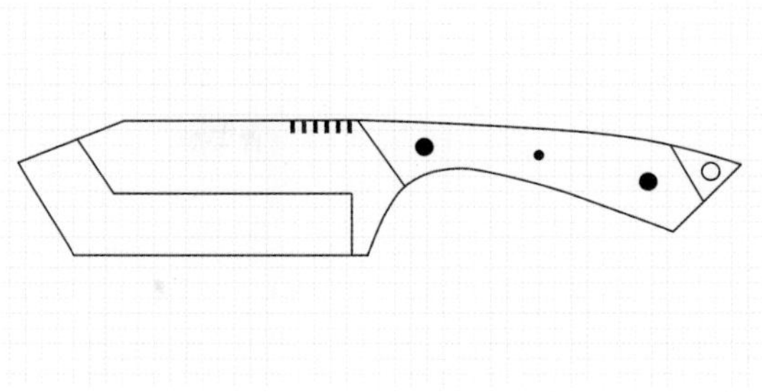

An example of a tanto design

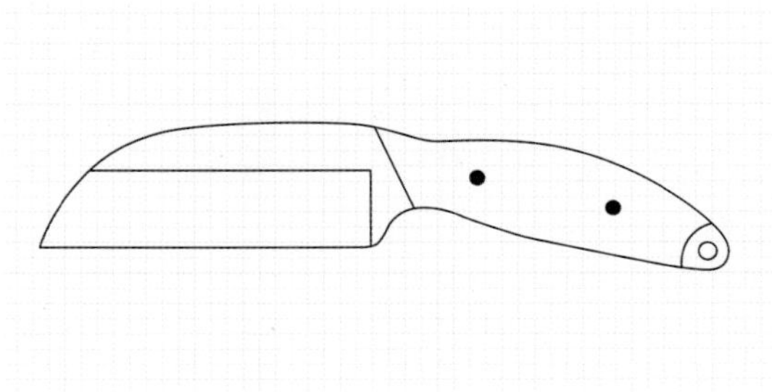

A knife with a sheepsfoot design

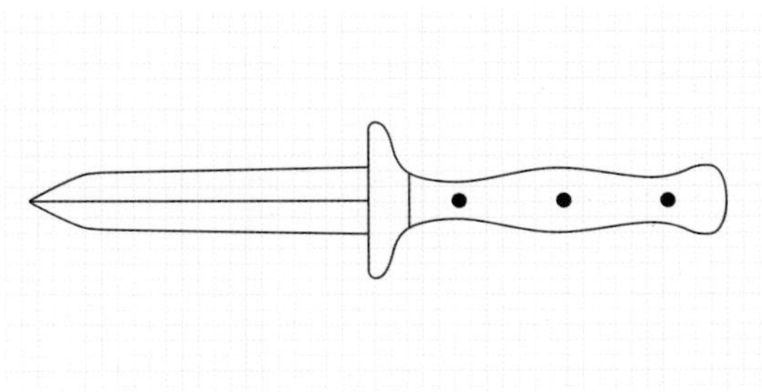

An example of a spearpoint design

Designing Your Knife

Now that you know your options, you have to figure out what you want your knife to do. Decide what you want the main purpose of your knife to be. While everyone wants to have the ultimate blade that is the absolute best at accomplishing every task, I don't believe this knife can exist. The qualities that make a knife the best for carving wood aren't necessarily the same qualities that make a knife the best for filleting sushi. While most blades will be multipurpose by nature, different designs will suit the knife to be better at some things than others. I don't see this as a limiting factor but, rather, try to use it to my advantage by recognizing the specific tasks I want my knife to excel at and tailoring the design around what will work to accomplish that goal.

A good start is to think about the knives you find most useful. Try to pinpoint what you like about them and why. Figure out what aspects of a particular blade work for you, and then think about what doesn't work. How can you incorporate the features you like into your own unique style of knife?

This gorgeous Damascus steel knife from Mike Mooney performs flawlessly in the kitchen but might not be the best choice for carving a primitive fire kit.

Function vs. Aesthetic

Making a functional knife doesn't mean it has to be unattractive. Your knife is a handmade piece of original art, and there is nothing wrong with treating it as such. While the aesthetic appeal of your knife shouldn't take away from its functionality, don't ignore this important aspect either. Take the time to design something you'll enjoy looking at.

While beauty is in the eye of the beholder, there are undoubtedly some common themes in people's perception of what makes a knife attractive. Curves can give eye appeal, so play not only with the shape of your blade but how it relates to the handle and its overall flow. Try to think of your knife as a whole while designing it instead of working on the blade and then attaching a handle to it. Most people will be turned off aesthetically by a knife that doesn't balance the length of the handle to the length of the blade. Having both be an equal size creates a knife that is pleasing to the eye and usually ends up feeling better in your hand as an added bonus. The exception to this is a bigger-bladed knife such as a machete or sword, which will obviously have a longer blade than handle by definition.

I suggest staying away from anything done for aesthetic reasons that will decrease the overall strength, integrity, and usefulness of the blade. One such example is a saw on the spine of a blade. While this makes the blade look really tough, such saws are useless without offset teeth. Instead, the teeth get caught up when you're trying to use the blade and make splitting wood by the process of batoning extremely difficult.

Beauty may be in the eye of the beholder, but there is a lot to be said about the beauty in simplicity. Personally, I find that an overly ornate and complicated blade can draw the eye away from the knife as a whole.

While being able to create your own design is an incredible outlet for creativity, make sure you design your knife for your equipment and skill level. It's important to ask if your design is even possible to make with your tools and experience. Leave the ornate brass knife guard and gem-encrusted pommel for when you have the experience of a few more basic blades under your belt.

Size Matters

Imagine trying to perform surgery with a machete or chop down a palm tree with a scalpel. Size clearly matters when it comes to the functionality of a knife, and bigger isn't necessarily better. It all comes down to the task you're performing. When thinking about the ideal size of your blade, try to imagine what limitations you could have by making it either too big or too small. How much of the blade are you actually going to be using? Will the length of the blade be sufficient to be useful? Will a big blade be too cumbersome? Will a small blade be able to handle the job? As a survivalist, a knife is no good to me unless I'm carrying it all the time. If it's too big, I won't always want to have it on me—and it might not be there when I need it. If it's too small, it might not be able to handle the function I need it to perform. It's important to find the balance somewhere in the middle.

While hunting big animals, most people assume the only knife that can do the job is a giant hunting knife. I prefer the opposite. As a minimalist who doesn't like to be weighed down while hunting, I often carry something as small as a scalpel to skin, gut, and quarter my animals in the field. The thin, tiny blade is easily worked between joints to break the animal down, and the razor-sharp edge makes all the cuts I need on the hide. Don't be limited by the ideas of others as to what makes a good blade for a job. You might be surprised by what actually works best for you.

The author butchering a moose. Try using a variety of knives to find out which kind you prefer for a specific task.

Make sure you also factor in how big your handle needs to be to make a comfortable fit in your hand. While the average knifemaker prefers a knife around 4½ inches (11 cm), I prefer a smaller knife handle to fit my small hands. There's not much worse than spending hours making a beautiful, functional knife only to realize that you don't enjoy using it.

Weighing these considerations will help you try to pinpoint the size of the knife you need. It can be hard to mentally visualize the exact ideal specifications of your knife, and everything looks bigger on paper. I recommend making a wooden blade template before you commit the idea to steel, so you can get a rough feel for whether or not your design will work for you.

The instructions in this book focus on helping you make a fixed-blade knife with a drop point design and full tang. This is my personal favorite style of blade for primitive survival, and it makes a good utility knife as well. Try making your first blade by following this build, and then as you revisit the process and become more comfortable with it, try experimenting to find your own style.

The project knife next to the original drawing of the design and the knife template

A beautiful blade made by Paul Brach with a pommel casting by Mark J. Hopper. Making an intricate blade like this takes a high level of skill and experience.

PRO TIPS: KEN ONION

LOCATION: KANEOHE, HAWAII

Ken Onion is an award-winning knifemaker and knife designer based in Hawaii. Ken was Kershaw's designer for ten years and has been a designer for CRKT for eight. In 2008, he was inducted into the *Blade* Magazine Cutlery Hall of Fame as its youngest member. Ken holds fifty-four patents and is the inventor of the assisted opening mechanism. He designed the popular "Ken Onion Edition Work Sharp Knife and Tool Sharpener" and has designed thousands of knives. Ken's work is known around the world, and he is a legend in the industry.

Knife style: "Diverse. Lately it's been much more retro, but prior to that it was organic."

How he got his start: "I first started making knives when I was a kid on the farm, modifying old bayonets and files. I got a copy of Knives Illustrated magazine at the drugstore and found a local guy that made knives, Stan Fujisaka. I hunted him down, and he taught me."

Best tip for a beginner: "Educate yourself. Read everything you can. There's an enormous wealth of information out there. If you're just gonna do it as a one- or two-time thing, start with basic tools, carbon steel, and a basic pattern. If you want to go into it in a larger fashion, buy the best equipment that you can afford so you only have to buy it one time."

On knife design: "Design your knife based on a story line. Think of a scenario for it to be used in and you'll never copy anyone else. Think in a theme, like a movie scenario. Create a futuristic story line or a jungle story line or a military story line. Design it off of a narrative. The best thing is to do it when there's no distractions. I like to design in the middle of the night when everybody's asleep, with one light on in the shop. My french curve's out, my pencil and eraser are ready to go, and I think of a story line before I sketch."

The very first knife I owned was designed by Ken Onion. His knife never left my side and ignited in me a spirit of adventure and confidence that influenced my life in huge ways. Much of my drive as a knifemaker comes not only from wanting to make my own knives but from the desire to make a blade that could someday inspire someone in the way that Ken's knife inspired me. I still carry one of Ken's blades, the CRKT Homefront, as my EDC (Everyday Carry) knife.

Making a Template

It's important to begin with an end in mind when you start building your first knife. Making a wooden template blade is a great way to see your design start to take shape and work out any changes you might want to make. By making a solid representation of your vision, you have something to reference and work toward as you make your knife.

I highly recommend purchasing a french curve from an arts and crafts store. This drawing template is very inexpensive and will help you to draw the curved lines of your blade.

MATERIALS AND EQUIPMENT

- thin plywood or thin hardboard piece
- graph paper and pencil
- spray adhesive
- scissors
- french curve
- band saw or saber saw
- wood file

1. Draw the design.

This drawing will eventually be your knife, so take your time and play around with the design. Use your french curve to help you make the curves of your blade. When you settle on a final design, make a photocopy to keep as a reference. Then, cut out the paper knife and glue it onto a piece of wood using spray adhesive.

2. Cut out the template.

Using a band saw or saber saw, carefully cut out the design on the piece of wood. Fine-tune the template by using a wood file to get rid of the saw marks and shape your knife by taking away everything up to the edge of the drawing. The more exact you are, the more you'll get a feel for what your knife will be like.

3. Critique the template and make any necessary adjustments.

Hold the template in your hand and really get a feel for it. Does the handle need to be longer or shorter? Is the blade length what you're looking for? Are the blade length and handle length the right proportions? If you have any doubts in the design, this is the time to fix it. It takes far less time to make a new template than to try to fix design flaws in your blade as you're working.

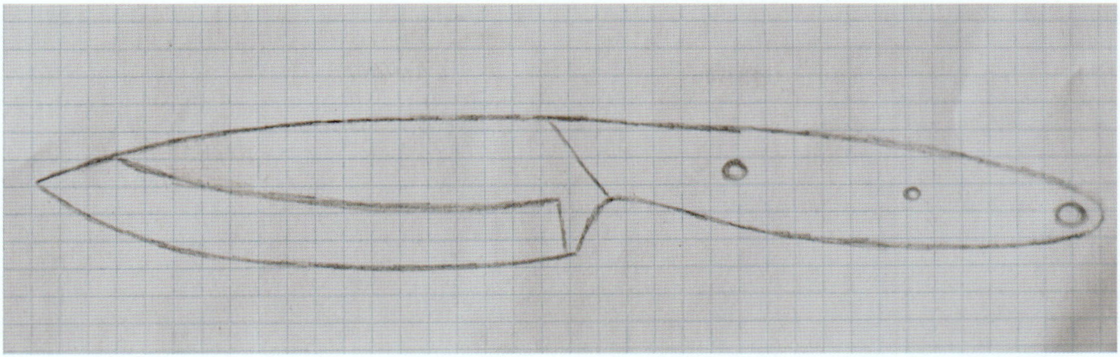

Drawing your design on graph paper isn't necessary but has a few advantages. Having a straight line to work from can be helpful, and the grid also helps give perspective in getting proportions right.

Though it may feel strange, try going through the motions of using your knife template. Time spent perfecting your design in the early stages can prevent you from having to troubleshoot problems later on.

1095

CHAPTER THREE

Understanding Steel

Steelmaking is one of the largest industries in the world today. Steel has countless uses in our daily lives, and there are many different kinds produced to meet these various needs. For the bladesmith, choosing an appropriate steel for a blade is just the first step. By manipulating its composition through a process of regulated heating and cooling, a skilled knifemaker works to bring out specific characteristics from the steel. The combination of these factors will determine the overall quality of the final knife.

Ultimately, knifemaking is a conversation between the bladesmith and a piece of metal. As a knifemaker, picking an appropriate steel for your knife will give you a good start. To bring out the qualities you want, you have to be able to tell the steel what you need from it. Understanding what's happening inside the chunk of metal you choose will help.

A variety of different kinds of steel stock ready to be made into blades

What Is Steel?

The base element of steel is iron. Iron is not only the most common element on earth but also one of the most common elements in the entire universe. Comprising much of our earth's core, it is generally found in the form of iron ore. This ore is heated through a process called smelting to create a nearly pure form of iron, known as wrought iron.

Like all metals, iron is crystalline in its solid state. These crystals are formed when metallic atoms bond. The crystalline structure of iron at room temperature, ferrite, allows the molecules to slip over one another easily. This makes it a soft and incredibly malleable metal. To make a lasting weapon or tool, a harder metal is required. Adding carbon to iron increases the hardness dramatically, creating an alloy we know as steel.

The first time steel appears in the historical record is about 4,000 years ago on the land mass that makes up modern-day Turkey. Ancient metal-workers would create usable chunks of iron by heating iron ore in charcoal fires. One of these smiths would have realized that by leaving the iron in the fire longer, the soft and ductile metal became much harder and stronger; the carbon from the charcoal was mixing with the iron to form a new alloy. While they might not have known the chemistry involved, they would have recognized the value of these properties and intentionally replicated the process. In this way, steel was discovered.

This wrought iron fighting ax, made by blacksmith Steve Ash, has a 1075 steel bit welded into the edge to make it a functional weapon. Without this piece of steel, the edge would be much softer. Many older tools used steel sparingly due to its limited availability.

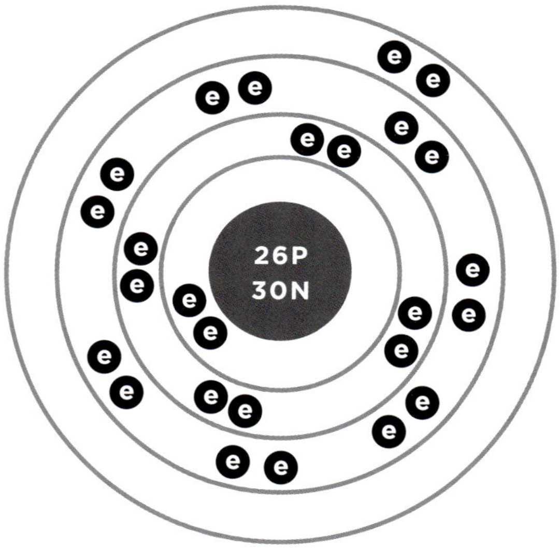

An iron atom is comprised of twenty-six protons, twenty-six electrons, and thirty neutrons, making it element number 26.

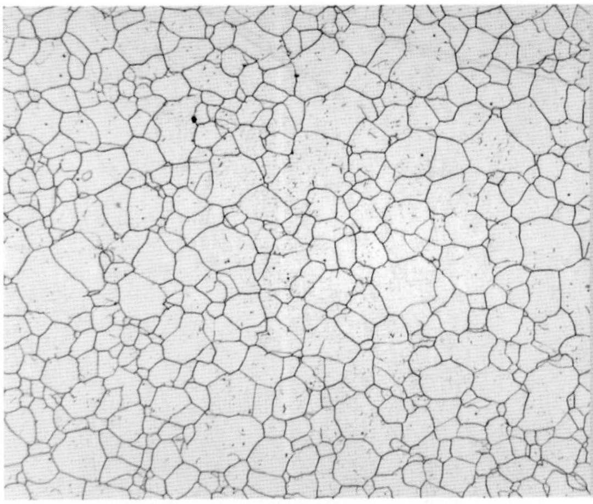

As metal solidifies, atoms arrange themselves into rows. This process happens in many different places all at once. Wherever these separate rows touch, boundaries are created. These create the "grains" in metal. Grain consistency is important to maintain when making a knife. If your grain size is inconsistent, the steel is prone to warping and non-uniform hardening.

The grain microstructure of steel. The visible lines are the grain boundaries.

Carbon atoms are much smaller than iron atoms. When added together, carbon arranges itself between the iron. The carbon atoms prevent the iron atoms from easily sliding over one another. In this way, the content and distribution of carbon helps to determine the hardness and strength of a steel.

In addition to carbon, most steel has a number of other elements mixed in. Each different element causes specific changes in that steel's molecular structure. By knowing the mixture of elements used to create a specific steel, you can predict how it will perform.

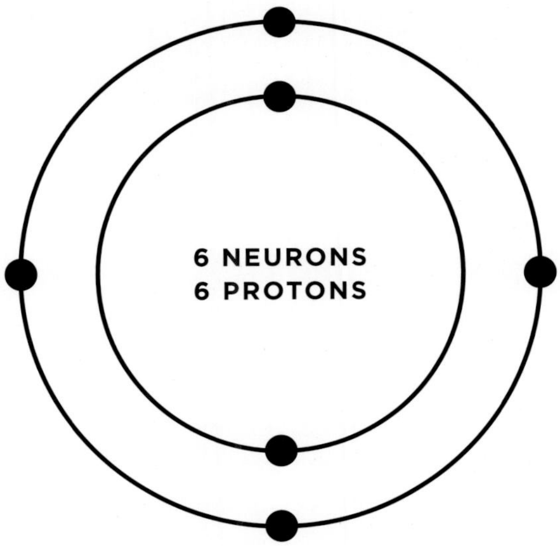

An atom of carbon only has six protons, six electrons, and six neutrons and is much smaller than an iron atom.

An alloy is metal formed by mixing two or more elements, at least one of which is metallic.

A rusty knife is something nobody wants to see. When iron atoms come into contact with oxygen in the presence of water, they give away some of their electrons in an oxidation reaction. We can tell that this reaction has happened by the presence of iron oxide, or rust. Given enough time, an entire piece of iron can turn into iron oxide.

The structure of the crystals ultimately gives each metal its specific properties. In addition to the changes caused by different elements, iron can exist in different crystal shapes with different temperatures. At room temperature, steel is a mixture of ferrite and cementite. By adding heat, this structure changes as the spacing between the iron atoms in the crystal changes.

This ability to assume different structures makes steel an incredibly versatile and useful metal. By purposefully manipulating the molecular structure through heating and cooling, you can bring out a number of different properties in the steel.

Survival cob used to chink a shelter wall. Adding straw turns dried mud into a highly functional material.

As a survivalist, I often find myself needing to change the qualities of a material to make it more useful for the purpose it's serving. One such example is making cob, otherwise known as survival cement. While dried mud by itself isn't very strong, by mixing in grass you completely change the properties of the material. The grass gives the mud something to hang onto, giving it structure and creating a much stronger final product. Cob can then be used to chink walls, build ovens, or even make durable bricks that can be used to build a structure. Adding carbon to iron has a similar result on a molecular level. Carbon gives the iron structure and allows for entirely different properties in the resulting metal.

Heat and Steel

There's an old adage in knifemaking, "Don't talk about religion, politics, or heat treating." You can talk to ten different knifemakers and hear twenty different processes they use to treat their metal.

Heat treating is the controlled heating and cooling of steel to cause specific changes in its properties. While "heat treating" is a blanket term for all of these processes, it is often used specifically in reference to hardening.

In any heat-treating process, ensuring uniform heating and cooling is important to avoid only affecting part of the steel. The exception to this is if you want different parts of the steel to have different properties, a process commonly used in making swords or other large blades.

Every metal undergoes specific changes at certain temperatures, known as critical temperatures. Variations in critical temperatures are what make different steels require different temperatures for heat treating. Do your research to find out the particular critical temperature for the steel you are using, as well as the recommended heat-treating processes.

The following is a brief overview of the different heat treatments you will use in knifemaking. By familiarizing yourself with how each treatment changes the molecular structure of the steel, you can understand the importance of each step. This will allow you to create the best possible heat-treatment routine.

Heat-treating a blade in the forge

Normalizing

When steel is heated above 1350°F (732°C), the carbon and other alloys start to dissolve in the iron. The crystal structure of iron begins to change into a new structure, austenite. The level of austenite increases as the temperature rises. When the steel reaches about 1450°F (788°C), all the crystal structures in the steel have become austenite. In this structure, the grain formation is broken up and formations are redistributed evenly.

When you cool this austenite steel slowly, a new structure called pearlite is formed. During the process of normalization, the steel is allowed to cool slowly in still air. This causes the metal to form a fine pearlite structure with relatively small, evenly distributed carbides. Large, unevenly distributed carbon can create a non-uniform structure in steel. This causes stress points in the metal, which can make for problems in the knifemaking process.

Normalizing is often used after forging, which can disrupt grain structure and cause internal stress. The uniform structure created during normalization evens out your steel and sets you up for success.

Annealing

The process of annealing is very similar to normalizing. The steel is brought to a temperature slightly above critical temperature but then undergoes a much slower and more controlled cooling. The slower austenitic steel is cooled, the more the carbon will be able to diffuse and pool away from the iron. This prevents the carbon from supporting the iron quite as much, resulting in a much softer metal. Utilizing this process before trying to drill or file your steel makes it much easier to work with.

Hardening

When austenitic steel is cooled very quickly, carbon doesn't have much time to diffuse and is trapped in its position. A new structure called martensite is formed. In martensitic steel, the carbon is distributed in a way that it holds the iron very rigidly in place. This process causes a lot of stress, and the resulting metal is extremely hard but very brittle. Using a hardened blade would result in a near certain breakage, so it's important to handle the steel gently until you can put it through the process of tempering to relieve some of this stress.

Quenching

Quenching is a process used to cool metal rapidly, usually using water or oil. Water produces an extremely quick quench, while oil cools the metal relatively slower. If cooled too quickly, certain kinds of steel can fracture. Preheating the liquid can also be used to create less of a shock and makes the oil less viscous so that it transfers heat more efficiently.

When steel is elevated to its critical temperature, it loses its magnetic properties. Metalworkers can gauge whether steel has reached its critical temperature without using a thermometer by testing it with a magnet. This point, around 1400°F (760°C), is known as the Curie point.

When grinding or working with power tools on your blade, the friction created can bring the temperature up to a point that causes changes in the steel. This is especially important after you treat your steel, as you can ruin your heat treatments. Keep an eye on the color of the steel. Straw-gold color is an indication that you are changing the tempered hardness. Blue means you have overheated and significantly softened your steel. Take it slow and quench frequently in water to make sure that your steel doesn't get too hot.

The blue color of this steel indicates it has gotten too hot, and the heat treatment has likely been affected.

Tempering

Gently heating steel causes the atoms in the metal to vibrate. In martensitic steel, this vibration allows some of the trapped carbon to escape and recombine. This process is called tempering and is used to take away some of the hardness caused by the hardening process. This brings the steel halfway back between annealing softness and being fully hardened. The steel will now be hard enough to be useful but soft enough to be sharpened and not easily break.

Quenching heated steel. It's important to hold the blade firmly and agitate it after it is immersed to avoid steam bubbles that could cause uneven cooling.

Selecting Steel

According to the World Steel Association, there are over 3,500 grades of steel. Trying to choose the best kind of steel seems like an impossible task. Many beginners start out using old files, tools, or whatever piece of steel they come across. While it's possible to make a knife out of some of these materials, it can be tricky to know exactly what you're working with. Most stock steel isn't expensive. It's worth spending a few dollars to know exactly what you're using so you can save yourself a lot of trouble. By picking a specific kind of steel to work, you can find out the exact steps you need to take to bring out the qualities in your steel that will make the best knife possible.

Common Steels for Knifemaking

The steels listed below are a starting point to get you familiar with some of the more common steels you'll find in the knifemaking world. This is by no means an all-inclusive list. Every type of steel has a separate system that gives it a specific number. It can be difficult to remember what the numbers of each system mean, but a quick Internet search can usually bring up everything you need to know. Certain additional elements, such as chromium and vanadium, have a big impact on the characteristics of the steel. One of the most important things to look for is the carbon content. The optimal range of carbon is between 0.6 percent and 1.6 percent. A steel near the higher end of this range is able to obtain a higher level of Rockwell hardness, and one toward the lower end of this range will be more resilient.

As a knifemaker, the main properties you are looking for in your steel are hardness and toughness. Your knife needs to be hard enough to hold an edge but tough enough to not break easily.

Know what you're working with before you throw any old chunk of metal in the forge. Certain metals, such as galvanized steel, can release toxic chemicals that can make you very sick. Often known as metal fume fever, this illness is characterized by flu-like symptoms. Full recovery can take one to three weeks.

Old farrier rasps are occasionally recommended for knifemaking. These rasps are often made of case-hardened steel, making them unsuitable for knives, as they can't be heat-treated properly. Avoid working with any unknown steel.

These two properties can work against each other, so finding the appropriate balance of hardness and toughness is an important consideration.

In addition to hardness and toughness, consider the qualities of forgeability, hardenability, edge retention, ductility, workability, corrosion resistance, availability, and finally, your own skill level. It's important to do your research on whatever steel you choose, as every steel performs differently. Each steel also has a slightly different heat-treating process to maximize its potential. Some steels are much easier to treat, and these are ideal to use as you're first starting out. The more you use a specific steel, the more you'll learn about how to treat it to bring out the qualities you want. However, don't let this stop you from experimenting with different kinds; the best steel for you might not be the first one you try.

A common system of numbering steels is the SAE steel system. The first digit is a code representing the main alloying element, the second digit represents the secondary alloying element, and the last two digits show the carbon content of the steel in hundredths of a percent by weight. For example, 1095 steel has only carbon (represented by the 1 and 0) with 0.95 percent carbon.

This bushcraft knife, made out of 1095 steel, has served me well on many expeditions in the field. This is a Saker bushcraft knife made by Abe Elias.

High Carbon Steels

High carbon steels are a group of steels that are primarily an alloy of steel and carbon, with very little additional elements added. The high carbon content gives these steels hardness and strength, and they tend to hold a great edge. These steels generally don't contain enough added elements to make them corrosion resistant, so rust can be an issue.

The group of steels known as "the 10 series" is extremely common for knifemaking. Perhaps the best known of these steels is 1095, which has a 0.95 percent carbon content. This steel is inexpensive, tough, and easy to find and holds a great edge. It is often seen with a coating on it to prevent rust. It is not necessarily a beginner's steel, as it can be a bit difficult to heat-treat properly; but overall it makes a great blade. In the 10 series, any steels between 1095 and 1045 would be suitable for knifemaking.

Both 1084 and 1080 are great steels to learn on. They are nearly identical, with only a very small difference in carbon content. These steels won't hold an edge forever, but they're easy to sharpen, easy to treat, and inexpensive. While they don't have the alloying elements that make other steels better performers, those elements also make them difficult to treat. These two are very forgiving steels and are a pleasure to forge. A 1084 steel is a great stepping stone in your knifemaking journey and what I recommend for anyone starting out.

Alloy Steels

Alloy steels used in knifemaking have a similar range of carbon content and are technically considered high carbon steels. However, this group of steels typically contains added elements to cause certain characteristics in the steel. These added elements can make certain kinds of alloy steels a bit trickier to heat-treat.

For example, 5160 is essentially 1060 plain carbon steel with added chromium. It contains between 0.56 percent and 0.64 percent carbon. While the added chromium isn't enough for rust-proofing, it does serve to strengthen the steel. This is a very forgiving steel that is great for forging. It also is fairly easy to heat-treat. It's tough, holds an edge, and does great if you're going to be hard on your knife.

Another type of alloy steel, 6150 was originally used in coil springs and has 0.48 percent to 0.53 percent carbon content and a small amount of added vanadium. This steel is nice to work with because it performs well even with less-than-ideal temperature control during the forging and heat-treating processes. It doesn't hold an edge quite as well as other steels but is very tough and easy to sharpen.

The high carbon content of 52100 steel (0.98 percent to 1.10 percent) makes it very hard, so it holds a great edge. It is very strong but can be a bit difficult to find in useful sizes. This is a good choice for hunting knives or other knives that you need to hold an edge.

A2 is a tough and relatively hard steel with 0.95 percent to 1.05 percent carbon content. It is very flexible but can be difficult to grind. It has less wear resistance than other steels but is often used in combat knives because of its toughness.

D2 steel has a 1.5 percent to 1.6 percent carbon content, so it is very hard. Its high 12 percent chromium content puts it just below the level of being considered a stainless steel, so it is resistant to rust. It is tougher than stainless steel, but not as tough as other tool steels. D2 has good edge retention, but it can be hard to sharpen. This steel can be difficult to work with and is considered a good steel for experts.

O-1 steel has a carbon content of 0.85 percent to 1.0 percent. It is a hard material with good edge retention, although it has a tendency to rust quickly. While it is relatively expensive, this is because it is usually sold as a precision ground bar, which makes for easy grinding. O-1 is a great beginner steel for using the stock removal method.

Stainless Steel

Stainless steel is characterized by its ability to resist corrosion. The presence of chromium in the steel creates a thin oxide layer on the surface of the steel, which helps to prevent the iron from oxidizing and creating rust. This makes it an ideal choice for knives that will be used in a wet or saltwater environment, or for a knife that won't be used a lot and is at risk of rusting in its sheath.

While stainless steel has a reputation for having poor edge retention and less strength than other kinds of steel, newer grades available have proved these theories wrong. When making a blade out of stainless steel, the stock removal method is preferred, as it can be unforgiving and difficult to work in the forge. The heat-treatment process can also be a bit challenging to do at home, so many knifemakers send their stainless steel blades out to be professionally done.

One of the most commonly known stainless steels is 440c. It has 0.95 percent to 1.20 percent carbon,

Stainless steel has the ability to hold an unrivaled mirror polish and can make an absolutely beautiful blade, as seen in this knife made by Paul Brach.

is wear resistant, and is a hard steel. It is nearly impossible to forge, but it is easier to grind than most other kinds of stainless. A knife made from 440c takes a nice edge that is easy to sharpen. While it takes a good deal of skill and patience, it does take a very nice polish. It is still used in some production knives but has been largely replaced by newer, higher-performing alloys.

CPM154CM is a stainless alloy that has 1.05 percent carbon. This steel was developed as a powdered metallurgy version of a different type of stainless called 154CM. The process of making CPM154CM involves using a metal powder and results in the uniform distribution of tiny carbides. This makes it a much better steel to grind and polish. It is also considered an improvement over 440c in edge holding, toughness, and corrosion resistance.

S35VN is a fantastic steel that was created specifically for the knife industry by Crucible steel in collaboration with renowned knifemaker Chris Reeve. While expensive and a bit hard to work with for a beginner, it has an excellent balance of toughness and edge retention. It is considered by many to be one of the best available knife steels.

Damascus Steel

I've long had a soft spot in my heart for the legendary beauty, strength, and flexibility of Damascus steel. Damascus has its origins in the wootz steel of Sri Lanka, where early steelmakers were able to reach high temperatures required for production in their furnaces by harnessing the power of monsoon winds. After it was introduced into the Arab world, the steel became famous and the favorite for making weapons. It was here that it was named after the capital city of Syria and developed its reputation as the best steel in the world. With its distinctive mottled pattern, superior edge retention, and excellent resistance to shattering, it reportedly had the ability to cut through other blades in battle.

More recently, the name "Damascus steel" has also been used to describe steel that is made by welding together layers of different kinds of steel into a billet. The billet is then forged, drawn out, and folded repeatedly. Though different than the original Damascus steel, the result is an incredibly beautiful and functional metal. The amount of time and effort required to make Damascus makes it much more expensive than most steels, and it can be difficult to heat-treat, depending on the combination of steels used in its creation. While it's probably best to invest in Damascus after you have some experience under your belt,

The term "tool steels" encompasses a variety of different carbon and alloy steels typically used for making tools. These steels are known for their superior hardness that will hold an edge even at high temperatures. D2 and A2 are examples of common tool steels.

it makes an incredibly beautiful blade and is my favorite type of steel.

Blade Thickness

Once you've chosen your steel, you'll have to plan for the thickness of your blade. As the thickness of the steel increases, the angle of the grind on your future knife changes. When you have more acute angles and less mass, your knife will have less drag and will cut easier. When choosing your blade's thickness, you want to have steel that is thick enough to be strong but not so thick that it adds unnecessary weight and makes the knife cumbersome. I recommend starting with a piece of steel between ⅛ inch (3 mm) and 5⁄32 inch (4 mm) thick for a good bushcraft knife.

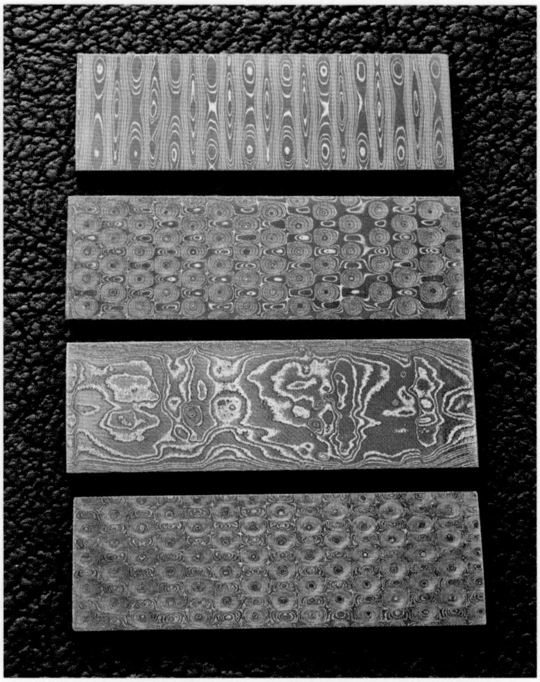

Damascus steel billets in a variety of different patterns made by Vegas Forge. The pattern in modern Damascus is determined by the way the steel is layered and pressed, and sometimes includes hundreds of layers.

Cutting Out a Blank

If you're forgoing blacksmithing and heading straight into stock removal, start with a piece of stock steel roughly the length that you would like your knife to be. You'll then need to cut out the rough profile of your knife in the stock steel using a band saw, hacksaw, or angle grinder.

If you plan on taking your knife to the forge, start with a piece of steel a bit longer than your knife template. This will allow room for error as you work. A lot of the basic profiling and shaping can be done with a hammer, but the learning curve of blacksmithing can make it difficult and time-consuming. If you're having a hard time with getting the right basic shape at the forge and you want to move on in the process, you can switch to stock removal at any time. After any forge work, the knife should be normalized and annealed as outlined at the end of chapter 4.

To cut out the profile, use a permanent marker or a soapstone pencil to trace your design on the steel blank. Use your angle grinder and a metal cutting wheel to cut off the excess metal. While the knife profile will end up a little bit bigger than you want, you can finish up the profiling using a sanding attachment on your grinder or a belt sander.

If you plan to forgo blacksmithing and head right to the grinder, you'll need to cut out the profile of the blade. Take care not to cut into any of the marked lines, as anything inside these lines will be a part of your blade.

PRO TIPS: ABE ELIAS

LOCATION: ONTARIO, CANADA

Abe Elias is a knife collector, reviewer, and custom knifemaker. As a bushcrafter and survivalist, his experience has given him a perspective of functionality that he brings into his design. He is a writer for numerous publications, including *Knives Illustrated* and *Blade* magazines, and is a licensed designer for a number of professional knife companies.

Knife style: "Practical knives with good ergonomics. I build knives that work."

How he got his start: "Knives have always been an unexplainable passion of mine, since I was a little kid. At several points during my life I tried to start knifemaking. Finally, in my mid-thirties, I got hooked up with a knifemaker who showed me how to make knives and it stuck. It was finally the right time for me to become a knifemaker."

Best tip for a beginner: "The best advice I ever got was from Jerry Hossom, and I can't thank him enough for it. That advice was to pick one pattern of a knife and stick with that one pattern until you learn to make it. A lot of what you do in knifemaking is muscle memory. In the event that you constantly adjust how you're moving to meet the different styles of knives, you're not learning anything. Practice on something that doesn't change so that the only thing changing is how you move. That's how you'll establish a set skill and see improvement."

On steel: "Start out cheap. Beginners get overwhelmed by steel, but it doesn't make sense to spend $80 on a piece of steel that you're learning on. Make the quality of your steel match your skill level. The two pillars of edge retention are the proper heat treatment as well as the proper edge geometry for the knife you're making. Make the most out of the steel you have."

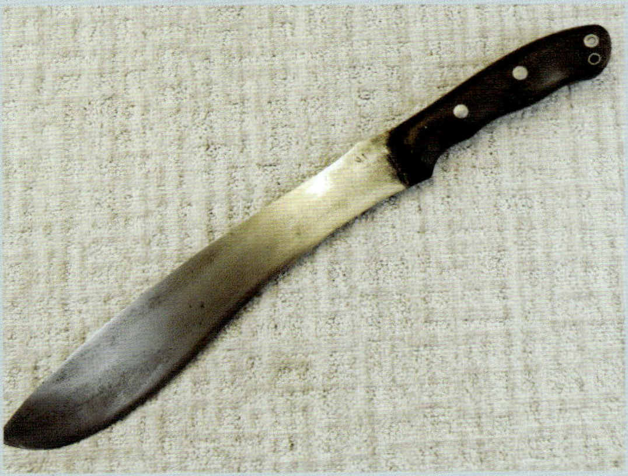

This Mako Shark, made by Abe, holds an edge unlike any knife I've owned and completely defied any expectations I could have had. When I asked Abe why his edge performed so well, he said, "I put my heart into it." This, combined with proper heat treatment of good quality steel (and an expert grind using proper blade geometry) created a legendary blade that has kept me alive on three different continents.

CHAPTER FOUR

Blacksmithing

Before modern-day grinders and power cutting tools, all metal knives were made through the art of blacksmithing. The local "smithy," as blacksmith shops are known, was at one point an integral part of every town. Today, most people have never set foot inside a blacksmith shop, but the prevalence of the surname Smith is a relic of how common this trade used to be.

I was first introduced to the art of blacksmithing while apprenticing as a farrier in my early twenties. I was constantly impressed by the master blacksmiths I met. They could seemingly communicate with metal and will it into any shape they desired with a few well-placed hammer blows. Mastering the art of blacksmithing is a lifetime pursuit, but by learning just a few simple techniques, you'll be able to start using this skill as part of your knifemaking process. This chapter is intended to give you some basic tools to start hammering.

Heating the Steel

You have your work space all set up, your design laid out, and a piece of steel ready to go. It's time to get that metal in the forge! Make sure all your tools are where you can have easy access to them, place your blade in the forge, and get it cranking.

I never quite understood the saying "Strike while the iron is hot" until I started blacksmithing. To properly work your blade, you'll need to take it out of the forge when it's at an appropriate temperature. While it's possible to buy a thermometer for a propane forge, most people gauge the approximate temperature and the resulting workability by its color. Steel needs to undergo changes during heating to become malleable. If you try to hit a piece of steel that's too cold, you'll only be working the outer layers of the metal and you'll get a mushrooming effect. In a worst-case scenario, you could even stress the steel to the point of breaking. It's also possible to get too high of a temperature, which is most likely when using a coal forge. Thin edges, and particularly the point, are most vulnerable. If using a coal-fired forge, take care to arrange your coals so that you can lay your steel across the fire evenly. If you shove your point deep into the coals and crank the air, it's possible to melt your tip completely off.

Keep an eye on your steel as it heats in the forge and wait for it to turn an orange color. Pay attention to how the color is changing when you start hammering. Once the steel dulls to a red, put it back in the fire to reheat. Make sure to keep watching the color of the steel, and never let it rest in the forge unattended. Extremely high temperatures can actually burn the carbon out of the surface layer of steel. If this happens, it's no longer high carbon steel and, therefore, can't be sharpened. Assuming the whole point of making a knife is to make a sharp cutting edge on a piece of steel, it's a game-over scenario for the piece of metal you're working on. While different steels can tolerate slightly different temperatures, I recommend erring on the side of caution. You don't need to take your steel to the upper reaches of what it can tolerate to hammer it into shape. I try to prevent my steel from turning a bright yellow to white color, and if it starts throwing a bunch of little sparks like fireworks, you're probably getting it too hot for your intended purpose.

This steel is ready to forge. Steel is forgeable between a hotter yellow color and a cooler red color. I recommend aiming for an orange color and working it until it cools to a red. While it can be worked when it is yellow, you will be pushing the ranges of appropriate forging temperature. It can be hard to push this boundary safely as a beginner, so play it safe to avoid accidentally overheating your steel.

Ambient light can affect how you perceive the color of the metal. For this reason, a lot of blacksmith shops are quite dark. Some people prefer to forge or heat-treat at night, as it can be easier to determine the color, and therefore temperature, of the blade. This also makes it easier to see if blade is evenly treated.

FAHRENHEIT	CELSIUS	THE COLOR OF THE STEEL	
2000°	1093°	Bright Yellow	Forging
1900°	1038°	Dark Yellow	Forging
1800°	982°	Orange Yellow	Forging
1700°	927°	Orange	Forging
1600°	871°	Orange Red	Forging
1500°	816°	Bright Red	Forging
1400°	760°	Red	Forging
1300°	704°	Medium Red	
1200°	649°	Dull Red	
1100°	593°	Slight Red	
1000°	538°	Very Slightly Red	
800°	427°	Dark Grey	Tempering
575°	302°	Blue	Tempering
540°	282°	Dark Purple	Tempering
520°	271°	Purple	Tempering
500°	260°	Brown/Purple	Tempering
480°	249°	Brown	Tempering
465°	241°	Dark Straw	Tempering
445°	229°	Light Straw	
390°	199°	Faint Straw	

FORGE WELDING

This color chart can be used to judge the relative temperature of the steel and its resulting workability.

Making a Game Plan

Before you start hammering, you need to come up with a game plan. Your steel will start to cool down immediately when you take it out of the forge, and every time you place it on the anvil, it rapidly draws heat away through conduction. This means you only have about six to eight good hammer strikes before you need to put it back in again to heat up. Use your time wisely! Take a good look at your steel and plan exactly where you'll be hitting it before you even put it in the forge. Make a mental plan of the steps you'll be taking so that you don't have to waste any time trying to figure that out when your blade is heated and ready to work. If you do have to stop hammering for a moment, hold the steel up instead of letting it rest on the anvil to prevent unnecessary cooling.

I like to make my plan by thinking of the entire process in terms of a series of smaller projects. Instead of being overwhelmed by the overall goal of making a knife, I think of the individual blows I'll use to start the tip or flatten the cutting edge. By focusing on each step as its own separate project, I can figure out a plan to most efficiently get from point A to point B.

Hammering

Now that you have your game plan, you're ready to get to work. Grab your steel from the forge and place it on the anvil, holding it securely in your nondominant hand with your tongs. Pick up your hammer and get in your first strikes on your steel. Congratulations! You're learning the ancient art of blacksmithing!

As you first start using the hammer, you'll likely find yourself wanting to choke up on the handle to improve your accuracy. This is a useful skill when you have to make precision blows, but if you're trying to move serious metal, it isn't an efficient way of striking. As you gain confidence, try to move your hand back to get the benefit of using the entire length of the handle. This will help to make your blows more potent and will put less stress on your body. Try to use the weight of your hammer to your advantage. If you find yourself holding on to your hammer with white knuckles, figure out how to get more comfortable. Find a rhythm and swing with your hammer instead of fighting against it.

You'll often see blacksmiths "tapping" the anvil after they strike the metal. This is a technique used by some smiths to keep the rhythm of their hammer, while giving them an extra second to consider where their next blow will land. This gentle tapping is much different than delivering the full force of a strike on the anvil face, which should be avoided as it can damage the face of the anvil.

Fire scale on an anvil. As you're working, you'll see small dark flakes coming off your steel. This is fire scale, and it will cause pock marks in your finished knife if you continue to work with it on the anvil. Take the time between firing your metal to brush the anvil clean.

Making a Hot Cut

While forging, you may come to a point where you realize a whole piece of metal needs to be cut off the steel stock that you're working. Instead of cooling down the metal and using an angle grinder or saw, you can do the work easily at the forge by using a hot cut hardy tool set into the hardy hole of your anvil. Take care to avoid hitting the hot cut hardy directly with your hammer as you can damage both your hammer and the hardy. If you don't have a proper anvil or a hardy cut, you can improvise this method using the edge of the anvil or a piece of steel with a ninety-degree edge.

MATERIALS AND EQUIPMENT

steel stock
hammer
tongs
anvil
forge
hot cut hardy

1. Heat up the steel.

Heat up your steel in the fire until it reaches a bright yellow color. Using your tongs, remove the blade from the fire and place the point you want to cut on the hot cut hardy tool.

2. Strike the steel over the hardy.

Use vertical hammer blows to strike the steel where it is lying on the hardy. When you see a dark line appearing where the hardy is contacting the steel, stop hammering immediately to avoid cutting all the way through. Make sure you don't hit your hammer into the hardy cut. I use a different hammer made from softer brass to avoid damaging my forging hammer. You can make a hammer for your hardy cuts by drilling a hole into a piece of copper or brass rod and inserting a wooden handle.

3. Bend the steel until it breaks.

Hang the cut end of the steel over the edge of the anvil. Strike the steel just beyond the cut, bending the steel at the breaking point. Keep flipping the steel over and striking the end to work the piece back and forth until it breaks.

The dark line indicating the hardy is almost through the steel.

Striking the steel over the edge of the anvil to break off the end

Knife Forging 101

Here's where you take of all the basic skills of blacksmithing and try your hand at putting them together to make a knife. It's important to remember that forging a knife is a skill that takes a lot of time and practice, so don't get frustrated if you have a hard time getting the metal to do exactly what you want. You don't have to be a master smith to make a great knife; in fact, some knifemakers never use the forge. Remember that the purpose of using the forge is to start shaping your blade so that you'll need to spend less time tediously grinding away metal. But your knifemaking process doesn't have to end here. Pace yourself while you're learning, relax, and enjoy the process.

MATERIALS AND EQUIPMENT

steel stock
tongs
hammer
anvil
forge

1. Hammer in the tip.

First, take a look at your steel and figure out where the tip of your blade will be. Heat the steel stock to working temperature. Hold it with the tongs on its side on the anvil, with the future cutting edge facing down. Hit the top corner at a forty-five-degree angle. As you strike, you'll notice the steel mushrooming out. Flip it on its side on the anvil and hammer it flat again. Keep repeating this process until you have what looks like the profile of the knife tip pounded in. This "tip" will actually change position as you continue to move the metal, moving from the bottom of the blade to the top. The natural tendency of the metal is to push things away as it gets thinner, so as you pound in the cutting edge, this "false tip" will actually end up rising to the top of the blade's profile.

2. Hammer in the cutting edge.

Look at your steel again and decide how long you want the cutting edge of your knife to be. Hold the blade over the edge of the anvil and hit the back of it with the hammer to make a small indented mark on the cutting edge side of the blade to designate the start of the handle. This will allow you to see the length of your future edge. Holding your steel flat on the anvil, pound along this edge. Flip the metal and hammer on the same part from the other side. This will cause the metal here to thin out, reducing the amount of material and starting the bevels of your knife. As you thin the edge, the tip will slowly move up into its proper place.

3. Maintain the flatness of the steel.

As you make progress in your blade, keep an eye on its overall shape. In addition to maintaining overall flatness, you'll have to work to keep the spine of your blade straight as well. As you notice it losing its straightness, flip the blade onto its edge so that the spine is on the anvil. Hammer into the cutting edge to correct this bend.

(continued)

Pounding in the tip of the blade. Note the direction of the applied force.

Work to thin the edge of the blade by striking along the cutting edge.

When steel is heated during the process of blacksmithing, the atoms vibrate out of their normal position. This allows them to easily move over other atoms in the alloy, and this is why it is possible to easily change the shape of the metal after heating it up in the forge. This movement and "scrambling" of atoms is also why it is so important to normalize and anneal the steel after forging.

Hammer in the tang to make it distinct from the business end of the blade.

A half-finished blade ready for the grinder. It's possible to move to stock removal at any point in the process. Before you shut off the forge, use your last firing to flatten the steel as much as possible before taking it through your heat treatments.

4. Define the tang.

Once you have a rough blade shape, shift your attention to the handle area, or tang. Hold the blade on its edge, with the spine on the anvil. Hammer in the indent that marks the start of the tang to further form this separation, again alternating with strikes along the side of the blade to correct the mushrooming effect of your blows.

> If you'd like to use your grinder to add in the bevels but still want to try your hand at forging, there is a quick and simple way to make your blade's tip. Using your hot cut hardy, cut off the tip of your steel at the correct angle. You can still try to shape your edge, but you don't have to worry about hammering so much that you get the tip into the right position.

5. Refine your work.

By now, you should be able to see the shape of your blade starting to form. Use the techniques you've learned to try to target problem areas and further refine the profile of your knife. There's no substitution for time and learning from your mistakes, so don't be afraid to make them. It can be helpful to give your body and mind some time to soak in what you've learned and head back to the forge for your next knife. If you've had enough for one blade, you can switch to stock removal at any point. Pull your knife out of the forge, let it cool, and cycle your steel through the treatments of normalizing and annealing to prep your blade for finishing it at the grinder.

Moving Steel

It's useful to think of the hot steel as clay. Imagine hitting clay with your hammer and what would happen to it as you delivered blows. While steel will be a lot harder to move than clay, the basic principles are the same. As you apply force, the malleable steel will move away from that force in the direction of least resistance. As you work, you will be adjusting where you are striking with your hammer as well as the direction of your strike. In this manner, you can control not only the direction of the force but where the steel will have a tendency to move by manipulating the path of least resistance.

As you work, the steel will eventually bend up at the ends, losing its flatness. It seems like common sense to want to hammer on the upturned ends to flatten the steel again, but this doesn't allow you to apply the right kind of force. By flipping the steel over and striking it in the middle, both ends will be pushed flat against the anvil as the middle moves away from the force of your hammer. Keep an eye on the steel as you work. Use the last strike or two to ensure your steel stays as flat as possible.

Pay attention to the direction of your hammer blows and the angle at which the face of the hammer strikes the hot steel. Small changes can make a big difference and will determine exactly where the force from your blow is directed and where the steel will move.

Flattening steel by using the hammer to apply force to the center of the opposite side

Normalizing

Before you move on in the knifemaking process, it's important to normalize your blade. The process of forging is very stressful on steel. The repeated cycles of heating and cooling, along with the physical rearrangement of the metal, wreaks havoc on the grain structure. Large or irregular grains can create weaknesses in your blade. This can lead to shattering or breaking during the process of making your knife—or later on while the blade is being used. By normalizing your steel, you can press the reset button on your steel's grain and ensure this doesn't happen.

To normalize 1084 steel, it is recommended to bring the steel through three different cycles of heating and cooling. Cycling the steel, at a lower temperature each time, will cause the grains to shrink further. After you let the steel cool completely from forging, heat it up to about 1500°F (816°C). If you don't have a thermometer, use your magnet to test the relative temperature of your steel. Since 1084 will no longer be magnetic when it reaches 1350°F (732°C), you can continually check the steel until you notice this change. As you're checking, take care to not leave the magnet in the heat for too long or it will lose its magnetism. Once you notice the steel no longer attracting the magnet, leave it in the heat for another twenty to thirty seconds so that the temperature rises to approximately 1500°F (816°C). Remove the blade from the heat and let it cool in still air.

Keeping a close eye on the blade in the forge

Once it is cool, heat it again to 1350°F (732°C). This will be the exact point at which the steel loses its magnetism. Take note of the color the steel is at this temperature. Remove the blade from the heat and let it air cool.

Your third cycle will only bring the steel to 1200°F (649°C), just before it loses its magnetism. Keep watching the steel and try to remember the color that it was in the previous cycle. You'll be looking for the steel to be a shade lighter than in the previous step. Estimate the steel's temperature as closely as you can and pull it out of the forge before it loses its magnetism. Let it air cool once more.

Bury the blade completely in your insulating material.

Annealing

After the steel has been normalized, it will be relatively hard. Without further processing, it will be very difficult to work with tools. By annealing the steel, you bring some of the softness back while maintaining the homogeneous grain structure.

Put the steel in the forge and heat it to 1500°F (816°C), or slightly after it loses its magnetism. Instead of pulling the steel out to air cool as in normalizing, you'll be cooling it at a very slow rate. The easiest way to accomplish this is by placing the blade into a container of insulating, fireproof material. Wood ash is easily obtainable and a great insulator. Another option frequently used by knifemakers is vermiculite. Vermiculite is a mineral often used in gardening and can be found at any store that sells gardening supplies. Allow the blade to cool overnight.

PRO TIPS: AUSTIN MCGLAUN

LOCATION: MINNEAPOLIS, MINNESOTA

Austin McGlaun is a custom knifemaker and a blacksmith. As a combat veteran of the 101st Airborne division, as well as a former police officer, he has a very real understanding of how necessary it is for a knife to perform well in the field. This experience and skill influences his style of knifemaking and knife design. Austin has designed knives for CRKT as part of the "Forged By War" program, and a percentage of the profits from the sale of his award-winning "Clever Girl" blade are donated to The Green Beret Foundation.

Knife style: "Highly functional knives with intent."

How he got his start: "My dad lost a knife that he liked to deer hunt with. He still had the leather sheath, so he decided to make a new one. He was a carpenter, and he used a D2 tool steel planer blade out of the planer in his woodshop and a deer antler for a handle. That started my passion for knifemaking, and I've been making knives ever since. I still carry that knife my dad made."

Best tip for a beginner: "Everyone is open to teach you. Ask a lot of questions so that you can learn from other people's mistakes. Becoming a professional knifemaker takes a lot of hard work and dedication to the craft. I've been making knives since the 1990s, and I still feel like a novice. There's a quote: 'The life so short, the craft so long to learn.' You will never stop learning. It's passion. Do it because you enjoy it."

On blacksmithing: "Never in the history of man has anyone hit a piece of metal until it's sharp. You'll always have to do some kind of grinding on it, even if it's just putting on an edge with a file. You might find some purists who think that if you don't do every step from point A to point B, you're not a blacksmith. It's not true. Blacksmithing isn't only about efficiency, and it should be enjoyable, so find what works for you. Do it because you enjoy it."

Austin's "Clever Girl" design for the CRKT Forged By War program

DEWALT

CHAPTER FIVE

Grinding

Now that your blacksmithing is done, it's time to start grinding. After the profile of your knife is refined, you'll be systematically removing material to create bevels. These bevels will form an angle that makes the cutting edge of your blade. This is the defining moment in which you really turn your piece of steel into a knife.

There's a variety of tools and techniques you can use to create these grinds, and your choice will depend on your own experience and preference. Professional knifemakers make this step look easy, but it takes a lot of practice to develop that level of comfort and skill. The trick is to go slow, be patient with yourself, and put in plenty of time behind the grinder.

Every new blade is another chance to practice perfecting your grinding skills.

Blade Profiling

Before you start making the bevels, make sure your blade profile is all set. If you skipped blacksmithing, you'll need to cut out the entire profile. Trace your template onto the stock steel. Remove any large pieces of steel outside your design with a hacksaw, band saw, or the cutting wheel of an angle grinder. If you worked your blade at the forge, correct anything you didn't accomplish in blacksmithing. Revisit the template and redraw your line, and remove any extra material. Use a belt grinder, files, or angle grinder to remove all material that won't be part of your knife's final shape.

As you get close to the exact outline of your knife, make sure to use a finer grit and move slowly to avoid taking too much. It's important to finalize your profile now, as it will determine your bevel grinds. If you haven't used a belt grinder before, this is a great time to get familiar with using one before you start making your grinds.

Make sure to completely finish your blade profile before you start grinding in your bevels.

Making a Scandinavian Grind

A Scandi grind is made by making a V-shaped bevel on the cutting edge of your knife. This bevel is brought up the side of the blade, stopping before the halfway point. It will follow the curve of the belly, keeping the same distance from the edge along the whole blade. The grind goes from the tip to a designated point on the ricasso.

During this part of the grinding process, you will only take the grinds to about 80 percent. If you grind the edge too thin, it will have a greater likelihood of warping or cracking during the quench.

MATERIALS AND EQUIPMENT

blade

belt sander

water quench

1. Mark your grinds.

First, find the centerline of your cutting edge. Color the entire edge of the steel with a permanent marker. Take a drill bit close to the thickness of your blade and run it along a flat surface next to the steel, dragging the tip along the blade edge. Flip the blade and do the same thing again. The line you create will give you a reference as to where your bevels should meet. Make a mark on the ricasso on both sides of the blade where you want your grinds to end to prevent you from taking that grind too far.

2. Grind off the ninety-degree angle of the steel's edge.

Hold the tang firmly with your nondominant hand, with the cutting edge side facing up. Using your dominant hand as a guide, grind the ninety-degree corner of the cutting edge into the belt to create a forty-five-degree angle. This will establish a straight and even line, which you will bring up further to create the bevel. It's important to stop before this grind reaches the centerline as you'll need to leave the edge at about the thickness of a quarter (or ⅛ inch [3 mm]). Repeat on the other side of the steel. I recommend using an old belt for this step, as the sharp corner of the steel will catch on the grit and ruin a new belt.

(continued)

Mark your blade to make a visible goal for your grinds.

The blade will naturally want to take the path of least resistance and find the flat spot against the belt. Use soft but steady hands to feel when the blade is resting on this sweet spot.

3. Find the flat spot.

Put a sharp, new 50-grit belt on the grinder. Fresh belts won't heat up super fast, so you won't have to stop as often. Using a soft hand, bring the steel toward the belt. Gently let the steel find the flat spot you've created and start moving the steel perpendicular to the grinder. You don't need to place a ton of pressure on the blade; just move it across and let the belt do its job. You'll use this technique of finding the flat spot to continue your grind every time you take your steel off the grinder.

4. Start working in your grinds.

Make light passes across the full length of the cutting edge, from the tip to just before the plunge line. Switch sides every few passes to keep the grind lines even. Some knifemakers start at the plunge line and work their way toward the tip. Other makers do the complete opposite. Try working the blade both ways and see which you prefer. Keep an even pressure on the blade, and keep the steel moving. Check your progress every few passes. Resist the urge to stop and check every few seconds, as this will have a tendency to give you a choppy line.

5. Keep an eye on the plunge lines.

Make sure you stop just shy of the plunge line, and don't get carried away and overrun it. The radius of the plunge lines will be set by using the flexible edge of the belt positioned over the side of the platen. You can start setting these lines, but don't cut them all the way in. It's easier to clean up your plunge lines with a finer belt after your blade goes through the heat treatments. Also make sure to take care grinding on the tip. There's less material to take away here, and it's possible to burn the tip up.

In between passes, dip your blade in the water quench to cool it down. Repeat this quench whenever the steel becomes too hot to handle.

6. Work your grinds up the blade.

Continue working your grind higher and higher up toward the spine. Every pass should be slightly higher than the last. If you notice your line becoming wavy, think about the amount of pressure you're putting on the blade and try to keep it consistent. If you find a specific area has less material taken off, try slowing down on those high spots and putting more pressure on the back of the blade. Work to remove about 70 percent of the total material that will make your bevel.

(continued)

PUSH STICKS

If you're having trouble putting pressure on your blade, consider using a push stick. A push stick is a flat piece of wood with a notch cut out to support the blade under the spine. This allows you to hold the blade against the belt with the stick instead of holding the blade directly with your hand. Push sticks can also be useful if you are having trouble holding on to the steel as it gets hot from grinding. Be careful when using a push stick after heat treatment for this same reason; if you can't feel the steel heating up, you could let it get too hot and ruin your heat treat.

The push stick in use

7. Clean up your grinds.

Switch out your 50 grit belt for a 120 grit and color in the ground surface with a permanent marker. Take your blade back to the grinder and start to clean up your grinds with the finer grit. Keep working until the marker is completely removed and then repeat the process with a 220 grit. This process of cleaning up the grind isn't only about the aesthetic of the blade but is a precautionary measure taken to prevent the blade from cracking or warping in the heat treat. Any deep grooves, scratches, or sharp edges will be susceptible to cracking in the quench due to the stress created by this process. As an added bonus, this will likely make the blade easier to clean up after the heat treat.

Some knifemakers use a jig to hold the knife in place at a designated angle and systematically grind away material. Jigs can give you great results and consistency on your grinds without much practice. If your goal is to make the most beautiful knife possible with the least time investment, a jig might be the way to go. However, if you want to fully experience the process of knifemaking, hand-grinding your bevels can be extremely rewarding. Developing your talent on the grinder also gives you greater versatility in the long run. Personally, I love the challenge of grinding my own bevels. Although it can be difficult, I enjoy the process and seeing how far I've come on each blade I grind.

While practicing my freehanded grinding at the Vegas Forge shop in Las Vegas, I suddenly found myself in a scenario I could have hardly dreamed up. I was minding my own business, working away, when suddenly expert knifemakers Ryan Weeks and Eric Ochs appeared by my side. As both men began to give me incredibly useful critiques on my technique, I realized that some of their tips were seemingly at odds. Slightly confused, I took a minute to watch each knifemaker work briefly on the belt grinder. Though they both make beautiful knives, I couldn't believe how different their techniques were from each other. Through their guidance, I was able to build my skill level and refine my technique in a way that took on aspects from both their styles. That day, I learned that there's not only more than one way to skin a cat; there's more than one way to make the knife that will skin the cat.

Pay attention to how far up the side of the blade your grinds are going, maintaining an even grind on both sides.

Grinds

There are several types of grinds that can be utilized to make the bevel on your blade. Each grind is best suited for different functions.

FULL FLAT GRIND

This grind is V-shaped and tapers consistently from the spine all the way to the edge. It creates a good balance of cutting ability and strength. While it is a very sharp grind, it can dull quickly but is easy to sharpen. This design is common in kitchen knives.

Scandinavian (or Scandi) Grind

Often favored in bushcraft knives, the Scandi grind is a flat grind that starts below the halfway point of the blade. By leaving a lot of material in the spine of the blade, this grind can maximize the durability of your knife. The lack of a secondary bevel means that the low angle will create a sharp edge. While the edge is not as tough as other grinds that offer a secondary bevel, it does make it very easy to sharpen in the field, even for a beginner. It is an excellent grind for carving. The location of the bevel makes it easy to see what you're doing, and in my experience, a blade with a Scandi grind will eat through wood with ease.

Sabre Grind

The sabre grind is a flat grind that starts halfway up the blade and, therefore, ends up having a bit thinner angle than a Scandinavian grind. While it isn't quite as good at carving, it tends to slice slightly better. Unlink the Scandi, the sabre grind typically has a secondary bevel.

Hollow Grind

In this grind, the bevels curve in to form a thin, very sharp edge. This edge tends not to be as durable as some other grinds, and it tends to need a lot of retouching to stay sharp. This edge can be slightly more challenging to initially grind as a knifemaker, but the edge created isn't difficult to resharpen. The incredibly sharp edge of a hollow grind makes it the grind of choice for straight razors and hunting knives. It tends to bind up at the top of the hollow when slicing through materials such as cardboard and isn't as well suited to being a utility knife as some other grind styles.

Convex (or Ax) Grind

A convex grind is a rounded grind that tapers to the edge. The mass behind the edge increases the durability of the edge, and it can be quite sharp. This grind is often used in axes, machetes, or choppers. The shape helps the wood separate in splitting, and the strength created by the shape of the grind makes it ideal for the heavy workload of these tools. This is a challenging grind and can be difficult to sharpen for a beginner.

Chisel Grind

In a chisel grind, one side of the cutting edge is flat, and the other has no bevel ground in. Because of the extremely low blade angle this creates, a chisel grind makes an incredibly sharp edge. This sharp angle also means the edge doesn't have great durability and needs to be constantly maintained. Chisel grinds are commonly used in food preparation as well as woodworking, as the bevel makes it easy to follow wood grain. It can be slightly inaccurate when slicing, due to the edge being off center. Knives made with this grind are often either right-handed blades or left-handed blades, depending on which side the bevel is on.

Asymmetrical Grind

An asymmetrical grind is created by utilizing two of these grind styles. A different bevel is then made on each side of the blade. A common type of an asymmetrical grind is the combination of a convex grind and a flat grind, known to make a very durable edge.

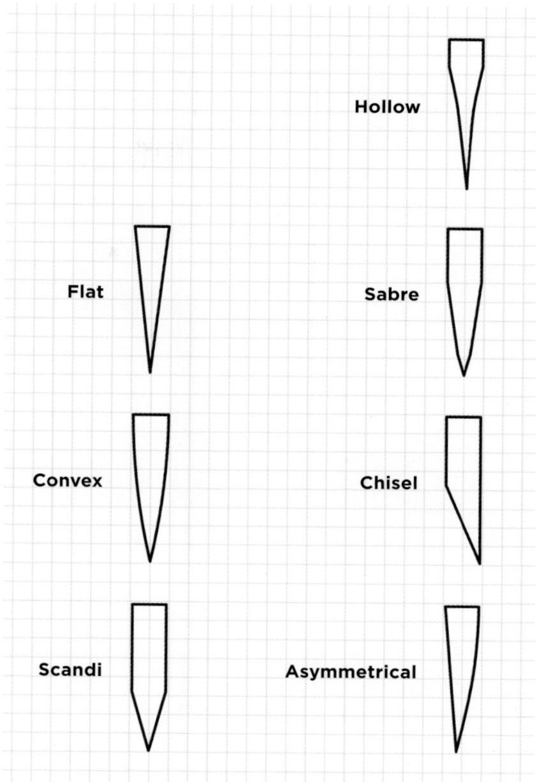

The different styles of grinds that will form your blade's bevel

The Scandinavian grind

Blade Geometry

A quality grind on a knife is produced not only with great technique but with a basic understanding of blade geometry. A good blade has an appropriate balance between overall strength, sharpness, and edge retention suited for its intended use. Unfortunately, there's no one-size-fits-all knife grind. Understanding what factors affect the performance of your blade will help you to choose the best grind for your blade and allow you to have a specific goal in mind.

Consider the function of your knife before you choose the grind. The abilities I demand out of my knives in the field are very different than those of knives I would use in the kitchen. A bushcraft knife would not be my ideal choice for finely filleting a good cut of meat, nor would I want to split wood with a delicate kitchen knife. The characteristics of the grind, combined with

COMMON BLADE ANGLES AND APPROPRIATE BLADE USE

In knifemaking, the angle of your cutting edge refers to half of the total angle created on the blade. For example, if you are putting a fifteen-degree angle on your knife, it will be fifteen degrees on each side, for a combined total angle of thirty degrees. These are some commonly used angles for different knife styles.

12° - 17°	scalpels, razors
18° - 25°	hunting
26° - 30°	bushcraft
31° and higher	machete, hatchet

the thickness and hardness of your steel, will determine what tasks your blade will be best suited for.

Generally speaking, a thinner knife will cut better than a thicker one. A thinner blade will have less drag and will pass through material easier. In losing thickness, however, you also lose overall blade durability. If you're rough on your knives, this could result in blade failure; and it doesn't matter how sharp your knife is if it's broken. Luckily, you don't have to decide between having a dull edge or a strong knife; the overall ability of your knife isn't set in stone by the steel thickness.

By manipulating the type of grind on the knife, you can work with the thickness of your steel to optimize the performance of your blade. The size of the angle created by the bevel will directly affect the performance of the edge. A larger angle will create a strong and durable edge, but it won't be as sharp as a low angle. A smaller angle will create a sharp blade, but it won't have enough metal behind it to make it durable. Decide what tasks your knife will perform and what compromises you're willing to make. By understanding this balance, you can figure out how big or small of an angle makes sense for your knife. For example, if you want an incredibly sharp edge but are worried about the strength of your blade, use a hollow grind on a thicker blade. This will create a small edge angle that will be incredibly sharp, while leaving a strong spine.

Secondary bevels are also used to affect the durability of a blade's edge. In addition to the primary bevel, a second bevel can be added to your original grind to create what is known as a compound bevel. This secondary bevel is usually more obtuse, which makes the edge itself more durable and less likely to turn. Additionally, the hardness of your steel can allow you to get away with more by reducing the probability of chips or a rolled edge. A softer steel, however, might demand a more forgiving grind.

As a survivalist, my knives have to take a lot of abuse. To me, a good bushcraft knife is one that can perform a variety of tasks at a moment's notice. From chopping wood to field dressing an animal, it needs to be able to take a beating while still having a useful edge. I love the edge retention, sharpness, and durability of a convex grind and definitely prefer them on my large chopping knives. For a smaller knife, I often use a style popular in the bushcraft community, the Scandinavian grind. It offers great control while woodworking, allows for a durable blade, and has a sharp edge that is easy to maintain in the field. While both make great blades, the convex grind requires a lot of skill and experience to execute and is a much tougher grind to learn on. I recommend starting with a Scandi grind for your first bushcraft knife.

TRIGONOMETRY AND GRINDS

I was never a huge fan of math, but if you're so inclined, you can use an equation to find your ideal angle. This is particularly useful if you'll use a jig and want an exact angle.

$\text{Degrees} = \text{atan}\,[(\text{thickness}/2)/\text{height}] \times (360/\varpi)$

If you search for a "grind angle calculator" online, you'll find a number of pages that allow you to input the variables for your specific piece of steel without having to do the math yourself.

Learning to Grind

A good grind on a blade is characterized by symmetry. With the exception of a purposefully asymmetrical grind, your grinds should match up as closely as possible on each side of the blade. The grind lines should be straight and not wavy, with an even thickness along the length of the grind and no dips in the bevel from overgrinding a particular spot. Another good indicator of grind symmetry can be seen by looking down at the edge of the blade. The area where the blade bevel ends on the cutting edge near the ricasso is called the plunge line. Symmetrical plunge lines are a sign of a good grind. This area is one of the last places you'll touch up as you grind, but keep in mind that you'll want them to line up.

If you've forged close to finish, you may find it useful to use files to finish out your grind. While this is a slow way to remove material, using files instead of power tools is a great way to really watch the progress you're making and make sure you don't take more than you want. If you're patient and enjoy this process, you can create the entire bevel on your blade this way. Expert knifemaker Wolfgang Loerchner makes incredibly beautiful blades from start to finish using files.

If you're comfortable using an angle grinder, you can clamp your knife to a stable surface and use a wheel to grind down the bevels. I don't have the experience or control with the grinder needed to create a decent bevel, and a more likely outcome for me would be a serious injury. If you feel confident in your angle grinding skills and don't have a belt grinder, go for it!

My preferred method is using a belt grinder. I find I have better control over my grinds with the ability to move the blade across a stationary surface. Make sure you have good lighting, and don't forget your safety equipment. The metal dust created while making your grinds is pretty substantial, so don't forget your respirator.

The Scandinavian grind. An example of a good plunge line on a blade by Abe Elias

Before you take your blade to the wheel, practice with something you haven't invested so much time into. It's possible to use a piece of wood, but if you have some steel around, it will give you a better feel for what grinding on your blade will be like and how much material will be removed. Developing skill on the grinder is all about developing your muscle memory. When you spend more time grinding, you develop an instinctual understanding of how your body needs to move to get the results you want. There is definitely a learning curve, so be patient with yourself. With a little bit of time, you won't need to think about your movements as much as when you're first starting out.

Developing your grinding style is about consistency, so eliminate as many different variables as possible. Stand with a slightly wide-set but comfortable stance to give yourself a stable base. Keep your elbows tucked against your sides and lock them into your hips. Instead of using your arms to move the blade, move from your core. Shift your weight steadily in your hips and think about using controlled and calculated movements. By working to create a pattern in your movement, you will find a comfortable rhythm that will make grinding much more predictable.

Pay attention while you're practicing to how the belt removes steel. Notice how a change of pressure and a slight angling of the metal will cause changes in where and how material is removed. Try changing where your belt sits on the grinder, adjusting it so the edge hangs slightly over the platen, or grinding plate. The amount of belt that is hanging over the edge determines whether you have a hard corner or a soft corner for grinding; the more belt that is over the edge of the platen, the softer your lines will be. Experiment with moving your steel over the edge of the belt to see how it cuts in differently. This skill will be especially useful when setting up your plunge lines.

Time behind the grinder practicing good, consistent form will help you learn how to make solid grinds.

CHAPTER SIX

Heat-Treating and Finishing

It's impossible to simply look at a blade and determine if it's been heat-treated properly. Through use, however, the contrast of a blade that has undergone the process and one that hasn't becomes obvious. A proper heat treatment will allow your knife to be the best it can be at the job it's intended to do.

The importance of researching your steel can't be stressed enough. Don't stick with the same heat-treating routine just because it worked for a different kind of steel. Each steel has its own unique requirements, and it's important to find and use the recommended process to get the most out of your steel.

While you may not be able to see the effects of the heat treat on a blade, the difference in performance is striking.

Final Tooling of the Steel

The next step of hardening your steel will make it much more difficult to work with tools. While you'll have to finish grinding your steel and polish it after it's been hardened, any work that can be done while the steel is still soft should be completed now. The handle scales will be attached with pins placed through holes in the steel, so drill these holes before moving on in your process.

In addition to the pins, epoxy is used to secure the handle on a full-tang knife. The purpose of the pins is to resist any shear force and reinforce the bond the epoxy creates. They also make setting the scales in place much easier. The number of holes, as well as their placement, is a matter of personal preference and aesthetic. Decide on the diameter of the pins you'd like to use, and use a drill bit approximately the same size or a bit bigger. If you don't have a bit slightly bigger, you can use your grinder to reduce the diameter of your pins so they fit in the holes. If you're worried about the weight of your handle and your knife being unbalanced, you can drill some extra holes in the tang to create a skeletonized effect. Remove any burrs you create with the drill bit, and sand down the sharp edges. If you have a countersink, slightly recessing these holes will help to remove the edge and ensure you don't have any weak spots for the heat treat.

Many bushcraft-style knives include a lanyard hole in the end of the handle. Depending on your preference, you can drill an additional hole for this purpose. I've found that looping a lanyard through the handle of a knife and adding a box stitch can be helpful in drawing your knife if the leather on your sheath comes up particularly high or if you have cold or bloody hands. Some people use a lanyard loop wrapped around their wrist to prevent dropping their knife, especially if they're using it over water or deep snow. I find this to be a pretty dangerous habit, due to the probability of the knife to then swing back into your body with force. There are ways to wrap a lanyard around your hand for a more secure attachment, but in my opinion, a proper handle design should allow you to easily grip your knife during use. If you're in a physical or mental state where you're worried about accidentally losing a grip on your knife, you probably shouldn't be using one. I've used knives in many of different situations, sometimes dealing with dehydration, exhaustion, and starvation. Even using blades in these compromising situations, I've never lost my grip and accidentally thrown one anywhere.

Drill any holes you'll need in your steel before the heat treat makes this process difficult.

Another feature to consider adding at this time is jimping. Jimping is a series of notches on the spine close to the handle. This textured area can create added grip, especially when choking up on the blade for detail work. Some people love it, and some people hate it. I've never really found jimping to be practical, but that's strictly my own personal preference. The series of notches should be evenly spaced, with care taken to keep them perpendicular to the blade edge. This can be a difficult task to achieve, so make sure you have the blade secured firmly before starting to work on it. The notches can be started by using the cutting wheel on a rotary tool, such as a Dremel tool. Then, you can use V-files, chainsaw files, and sandpaper to smooth them out and remove any harsh edges. After the heat treat, they can be finished and polished along with the rest of the blade.

Before you move on to hardening, take one last good look at your blade to check for any potential weak spots that need to smoothed and rounded. During the heat treat, you'll be focusing on three main variables: heating, cooling, and time. Make sure you have done all your research to find the appropriate mix of these variables for your steel.

Adding a lanyard hole, jimping, or both are optional steps to customize your blade.

The Rockwell Hardness "C" Scale is a scale that is used to quantify the relative hardness of hardened steel. This scale is based on the material's indentation hardness, or how much a material will indent when a force is applied to it. By measuring the depth of the indent from an applied force and then using this data in an equation, a rating on the Rockwell Hardness Scale is assigned. While this test requires costly equipment and is, therefore, not practical to perform at home, it can be a useful system to be familiar with when looking up information about the heat treat for the specific steel you're working with. Most blades will fall between HRC 55 and 66, with 55 being a softer steel and 66 being on the upper end of an acceptable blade hardness.

Hardening

Hardening is the first step of the final heat-treating process. This is accomplished by bringing the steel to a high temperature and then cooling it down very fast. Your steel, which is still soft and workable from annealing, will become extremely hard. This hardening is necessary to make sure your blade can hold an edge. While hardening, it's important to keep a close eye on the steel to avoid letting any part of your blade get too hot. Speaking from personal experience, it can be heartbreaking to spend so much time and effort on a blade only to ruin it this far into the game. Additionally, the crystallization that happens when steel goes through the process of hardening makes it extremely brittle. Be very careful with your hardened blade until you toughen it up through tempering. Dropping a hardened blade can make it shatter like glass. Don't start the hardening process unless you know you have time to temper it immediately afterward. I've heard horror stories of people leaving their freshly hardened blade on the workbench overnight to finish in the morning, only to find that it cracked overnight.

Hardening your blade in a pit fire is a great alternative to using a forge.

If you made a two-brick forge, you can heat-treat your blade by following the process below, substituting the forge for the pit fire. By holding the blade in your tongs and moving it around in the forge, you can keep an eye on how hot different parts of the blade are getting. It's also very easy to move the blade in and out of the forge, checking it with your magnet.

If you didn't make a forge, you can achieve the same results by heat-treating your blade using a pit fire and a hair dryer. I started using this method after watching a YouTube video made by Walter Sorrells. Being on the road frequently and not always having the right equipment, I love the ingenuity, simplicity, and ease of being able to use any existing fire pit to heat-treat steel.

It's quick and simple to make a fire pit if you don't already have one set up. Be sure you line your pit with a ring of rocks or bricks to prevent the flames from spreading, and have a water source nearby in case a strong wind causes fire to leap the ring. It's always a good idea to check with local officials to find out the regulations regarding fire and your area, and obtain a permit if necessary. You'll be using hardwood charcoal instead of regular firewood in your pit. This causes hot, dense coals that will get your steel up to temperature quickly and easily.

Your fire will need an air source to create a hot spot in the fire to heat your steel. Attach a piece of metal pipe to the end of a hair dryer with fireproof, metal tape. The steel pipe will be positioned in the center of coals, and when the hair dryer is turned on, the airflow will stoke the coals in that area. Make sure to avoid galvanized pipe because of the toxic fumes it will release. You can use a variety of different creative air sources, but a hair dryer is cheap, readily available, and sufficient.

It's also possible to estimate the temperature of your steel by sprinkling salt on it. Salt melts at 1475°F (802°C), so if the salt melts, you can guess the temperature is at least this high. Remember that this will only account for the surface temperature of the steel, and you need your blade to have a uniform temperature throughout to get fully hardened.

A process called differential hardening is often used on a knife's edge to make that part of the blade harder than the rest of the steel. By putting clay or a similar material on the spine of a knife, you can harden just the edge of the blade. This creates a knife that will hold an edge while maintaining a soft, flexible spine. While this may not be necessary for every blade, it is especially useful in making swords and larger knives more durable.

The marking on this blade, called the hamon, is a visual example of differential hardening. The temper line distinguishes between the soft spine and the hard cutting edge. This blade was made by Ryan Weeks.

Quenching

Quenching is the process of rapidly bringing the temperature of steel to room temperature after it has been heated. Quenching steel after hardening "sets" the crystalline structure, preventing it from going through the changes it otherwise would if allowed to cool slowly. The medium used to quench the steel, or quenchant, determines the speed of cooling. Saltwater would create an incredibly fast quench, regular water slightly less quickly, and oil slower still. The faster the quench, the harder the steel gets. However, the process is very stressful to steel. A very fast quench could result in cracking or warping. If the steel is quenched too slowly, it won't be fully hardened.

Different types of steel can have different requirements for the quench process. For example, 1080 steel requires a fast-quenching oil such as Parks 50, as you have about one second to cool the steel from around 1500°F (816°C) to 900°F (482°C). Professional-grade quenchants were developed by engineers to remove heat from steel at prescribed rates. The use of a commercial oil quenchant will allow you to get the most from this alloy, as you will increase the chances of getting full transformation from austenite to martensite without creating pearlite. If it's not available, almost any kind of oil can be used, from motor oil to cooking oil. Canola oil heated to 120°F (49°C) does a great job and is widely accepted as a good quenchant for 1080. Preheating the oil before you quench not only makes the oil less of a shock for the steel, but it creates a thinner liquid to more quickly disperse heat. Make sure you don't heat the oil over an open flame, as this can be a fire hazard. A small metal quench container is easily heated using a hot plate.

Before you quench, make sure you have a plan of action in case of a flare-up when you introduce the hot metal to the oil. Flare-ups are caused by the hot metal igniting the vaporized oil on the surface and are usually brief but can be alarming if you're not expecting it. Submerging your blade completely in the oil can help prevent this from happening. Keep a fireproof lid nearby that can be dropped on top of your quench container to cut off the flow of oxygen in case of a problem. It's good practice to also have a fire extinguisher handy. Make sure you don't try to use a hose, as water can spread an oil fire. Wear gloves to protect your hands against any possible flames, and don't stand with your face directly over the quench.

Hardening in a Pit Fire

MATERIALS AND EQUIPMENT

tongs

forge or charcoal pit fire

charcoal

quench tank with preheated vegetable oil

telescoping magnet

1. Get the fire going.

Start the fire and make sure there is a good bed of coals. There needs to be enough fuel to keep the whole blade covered during the hardening process so that it heats up as evenly as possible. You may need to add charcoal as you go to keep up the coal bed as the fire burns down, so keep it close by.

2. Place the blade in the fire.

Position the blade so that it's in the hottest part of the fire, right where the air hits the charcoal. Place the tip just outside of that hot spot. The tip and the edge are the thinnest parts and will heat the quickest. By keeping them just outside of the hottest part of the fire, it is less likely they will overheat.

(continued)

Place the blade in the hot spot, with the tip just outside of it.

Watch the color of the blade as it starts to heat up, aiming for consistency.

3. Move the blade around.

Keep moving the blade around in the bed of coals to prevent it from overheating in any one spot. As you move the blade, check for a relative consistency of color in the metal. While it's hard to tell the exact temperature based off of color alone, this tells you whether or not the metal is heating up evenly. While the edge of the blade needs to be hardened, the tang won't need this change to perform its job well. If your tang isn't getting heated as much because you're holding onto it, don't worry about it.

4. Test the blade with a magnet.

As the blade begins to heat up, continually check it with a telescoping magnet. Metal starts to lose its magnetism at about 1425°F (774°C), so when the blade stops attracting the magnet, you're almost there. It's almost impossible to hit an exact temperature without a thermometer, but by estimating the temperature, you can get very close. Keep the blade heating for another fifteen to twenty seconds after losing magnetism, and you should be at roughly 1500°F (816°C), the ideal temperature for hardening 1080 steel.

Quench the blade quickly in preheated oil.

Use the file on the side of the blade to test for the steel's hardness.

5. Quench the blade in oil.

Once you think your blade is up to temperature, grab it by the handle with the tongs and immediately plunge it into the oil quench. Place the entire blade, tip first, in the oil and move it around. A vapor barrier is formed around the steel when it is submerged in the oil and can act to insulate the steel. This agitation helps the blade to cool quickly and evenly to prevent spotty hardening. Try to move the blade from spine to edge as opposed to side to side, to reduce the likelihood of warping.

6. Test the blade hardness.

When the blade is cool to the touch, remove it from the oil. Use the edge of a file to test the side of the blade. The file should move easily across the surface without digging in. If you can scratch the surface with the file, the blade didn't reach a high enough temperature and it will need to be heated again.

Tempering

After you finish hardening your blade, wait until it cools down to room temperature. You can then carefully use soap and water to clean off the oil from the quench. To temper, steel is brought up to an elevated temperature well below the steel's critical temperature and kept there until it is uniformly heated. After soaking in the heat, the steel is air cooled. This cycle is then repeated to make sure the steel is fully tempered. After you complete the temper, the heat treatment of your blade will be completed. The edge should be hard, yet tough, and your knife is one step closer to completion.

If you don't want to heat-treat your own blade, you can send it out to be professionally heat-treated. This can be a safe choice if you're using a steel that requires a particularly complex heat treat, or one that is above your ability level. Most companies or individuals that provide this service require a minimum amount of blades to complete an order, so you might have to wait until you have several blades ready to go before you send them out. Sending your steel out to be heat-treated will require a small investment, but it saves you work and you won't have to worry about screwing up a blade after you put all that time into it.

Multiple tempering can improve the toughness of your steel.

Tempering 1080 Steel

MATERIALS AND EQUIPMENT

blade

oven

tongs

1. Preheat the oven.

Preheat the oven on bake to a temperature of 400°F (204°C). The oven needs to reach a stable temperature before you start the tempering process. Depending on your oven, this will probably take between fifteen and twenty minutes. I like to turn the oven on before I start the hardening process so that it's good to go.

2. Let the blade heat up.

Put your blade in the oven, directly on the oven rack. Leave your blade here for about two hours. Your blade needs to soak up all the heat and reach a uniform temperature.

3. Remove the blade and let it cool.

Take your blade out of the oven and set it somewhere to cool down completely.

4. Repeat.

Put the blade back in the oven and repeat the process once more. For 1080 steel, two cycles should be enough; but some steels require more. When the blade is finished, pull it out, let it air cool once more, and you're good to go!

Grinding the Surface

It's time to head back to the grinder to clean off any scale that formed and finish the rest of the grinds you started. At this point, it is especially important to not let your blade get hot on the grinder and ruin the heat treat. Have your water quench ready and dip the steel frequently. I never wear gloves after this point so that I can feel the steel heating up and be reminded to cool it down. If your steel is changing colors, you're letting it get too hot. If you see a straw-gold color, you are changing the tempered hardness. If it gets blue, it is overheated and you've significantly softened your steel.

Depending on how much scale your blade has accumulated, you might have to use a coarse belt to clean it off. If you're noticing the steel heating up quickly and the scale isn't coming off easily, you can switch to files to remove it.

An inexpensive and useful addition to your shop is a hundred-pound lift capacity magnet. While not necessary, it allows you to hold the blade flat and parallel to the belt so that you can grind the entire side at once. Be VERY CAREFUL, as you will no longer be holding the blade and can't feel how hot it is getting. Don't grind for more than a few seconds before checking and quenching your blade.

Once the scale has been removed, it's time to finish off your bevels. You'll notice that the steel is slightly more difficult to work now that it's been hardened. Between this harder steel and the need for frequent quenching, it can be slow going. This can work to your advantage, as it forces you to take your time and you're, therefore, less likely to make mistakes as you put the finishing touches on your grind. Finish up your plunge lines, making a crisp edge on the cuts.

Grinding the blade with the use of a magnet. This also allows you to make sure the side of the tang is completely flat in preparation for attaching the handle.

After you've removed the bulk of the material, don't forget to transition through the finer grits to clean up your grinds. Don't worry about sanding down the tang, as it will be roughed up before you put the handle on. Take your blade down to 400 grit with the belt before switching to hand sanding.

Polishing

Polishing your blade is about more than just making it look pretty. A highly polished knife creates less friction and will cut better. A polish on your knife can make the difference between an adequately sharp blade and a wickedly sharp one.

When polishing, don't waste your time touching up the tang. The handle scales of your blade will be affixed here with epoxy and will be abraded to create a better point of contact for the adhesive. Your focus should be on the part of the blade that will be exposed after this happens, so take the polish just past where your handle will start.

Use a sandpaper that is made for sanding metal. Silicon carbide, also known as wet/dry sandpaper, is a good choice for especially hard steels. I find aluminum oxide to be sufficient in most cases, and it tends to last longer. Lubrication on your steel can also make your paper last longer as well as help prevent binding and clogging. With the coarser grits, use an oil such as 3-in-1 or WD-40. As you get to finer grits, glass cleaner works great.

Start with the blade clamped firmly by the tang to a stable surface, with the point facing you. Take a look at your last grind lines and start sanding at a forty-five-degree angle to these lines. This will allow you to distinguish between your old grinds and your new, finer ones.

As you polish, you'll be switching to finer and finer sandpaper. Focus on removing the grind lines from the grit just above the one you're using. Once these lines have been removed, you'll switch to a finer grit and change your angle of sanding again. Make sure you have good lighting for this process so you can see the lines of the previous grind.

Polishing takes a bit of time and elbow grease, and your hands will get surprisingly tired and sore. Instead of holding the paper in your bare hands, try wrapping it around another object to make a sanding stick. You can use wood, steel, leather, or whatever material you have lying around. Adhering a foam or leather pad to a hard surface before wrapping it can provide some give, making it form around the steel better and making it less likely the paper will rip.

The polished blade covered in protective tape. Painter's tape won't leave a sticky residue on the surface of your steel.

You'll be doing the final sharpening of your edge when you finish all the work on your blade, but make sure you get the grind lines out now. It can be especially difficult to polish around the ricasso after the handle is on. Keep working up in grit until you get the polish you want. While 1000 grit is usually sufficient, some people take it up to 2000 grit or even higher. The level to which you take this is completely up to you. The higher you go, the more of a "mirror" finish you will have; you can actually see your reflection in some finely finished blades.

Once you finish, cover your blade completely with painter's tape. This will not only protect you from accidentally cutting yourself, but it will protect it from getting scratched as you continue your work.

PRO TIPS: KAILA CUMINGS

LOCATION: TROY, NEW HAMPSHIRE

Kaila is a custom knifemaker and popular knife reviewer on YouTube. Though many of her designs have a tactical feel, her skill as both a hunter and a survivalist has given her a perspective rooted in real-world experience. Her dedication and hard work have paid off, as she is now one of the most well-known and talented women in the industry. Kaila makes all of her own knives from start to finish and has released a collaboration design with Pohl Force.

Knife style: "Tactical, yet practical."

How she got her start: "I started by doing knife reviews on YouTube. Eventually, I decided I wanted to start making my own. I had never done it and didn't know anyone who made knives, so I started watching instructional videos to give me an idea of where to start. I spent a ton of time in the shop learning how to use the tools and how to work with different materials and eventually figured it out."

Best tip for a beginner: "Be patient with yourself. A lot of the skills (in knifemaking) take a ton of shop time to really get the hang of. If you get frustrated or stuck, try looking up videos online to watch how different knifemakers do it. If you don't know anyone who makes knives, the Internet is a great resource."

On finishing your blade: "I buy PCB acid and make an acid bath for etching. After the blade is finished, I put a design on it with nail polish. When you soak the blade in the acid, it takes away material from everywhere but where the nail polish is. You can really get creative and personalize your knife by making different patterns or writing words. It adds something really unique to each knife."

One of Kaila's blades

CHAPTER SEVEN

Making a Handle

Now that the majority of the metalwork is done, it's time to switch mediums. There are countless options for the combination of materials that can be used to build a knife's handle, and each will create a different look. The handle aesthetic is a defining part of any knife. The materials you choose will help to create the unique character of your knife.

One of my favorite parts about making my own knife is getting to create my own perfectly fitted handle. Most knives are made to have a one-size-fits-all grip, despite the fact that hand shapes and sizes can vary greatly. By carefully shaping the material, you can create a grip on your blade that is personalized to fit the shape of your hand exactly.

A good handle on a blade can turn a good knife into a great knife.

Scale Materials

To make the handle for your blade, you will attach two pieces of material to the outside of your tang. These pieces are called scales. You can make your scales out of a wide variety of natural and man-made materials. While there are some benefits and drawbacks of using certain materials, a lot of this comes down to personal preference.

When choosing the material for a knife handle, you should take into consideration the environment and kind of abuse your handle will need to take. If you're going to be hammering on your handle frequently during the process of batoning, it might not make sense to use a softwood that could be easily damaged. Changes in temperature and humidity will also make some natural materials shrink and swell, which could affect the integrity of your handle.

The following is a list of common handle materials for fixed-blade knives. In experimenting with different handle materials, be creative but don't let it be at the expense of function. Regardless of your choice in material, the most important part of selecting your scales is to make sure to have equally sliced, perfectly flat halves.

Wood

The availability of a great variety of different kinds of wood allows for a wide range of choices in color and design. These tend to make more durable handles. Knife scales are often made with stabilized wood. During the stabilization process, wood is injected with a polymer resin. This makes it more durable and less likely to crack or warp, especially when transitioning through different climates. Some woods definitely need to be stabilized to perform well, but very hard or oily woods don't need to undergo this process.

Antler, Horn, or Bone

These materials give a beautiful, traditional look to a handle. As a primitive survivalist, I love the idea of using the same materials as my ancestors to make a knife handle. However, they don't naturally provide a great grip and can be slippery, especially when there's blood on the handle. They are also affected by temperature and moisture and are prone to cracking.

G10

G10 is a composite made from fiberglass cloth and resin. It is tough, lightweight, and makes a durable handle. G10 can be slippery to hang on to when it is wet.

Micarta

Micarta is a synthetic material made out of a fabric, such as canvas or linen, soaked in resin. It performs similarly to G10 but is nicer looking and a bit more expensive. Users agree that G10 is slightly grippier when the handle is dry, but Micarta is grippier when wet. When the handle is exposed to an oily or greasy liquid, however, it will make the Micarta a bit slippery.

Carbon Fiber

Carbon fiber is made from woven strands of carbon set in resin. Lightweight and expensive, the woven strands catch the light, making a beautiful handle. While carbon fiber is known for its strength, it can also crack and come apart from an impact, such as dropping your knife.

Wooden scales can be purchased from knife supply companies. These scales come as a matched set.

In many ways, the material of your handle is only limited by your imagination and ability to creatively improvise. When Kelly McGuire Pilcher had to have her first rib removed for medical reasons, she was able to take it home with her after surgery. Her father suggested using it to make a knife handle. Texas-based knifemaker Travis Payne was the man for the job and was able to create this knife using resin and Kelly's rib.

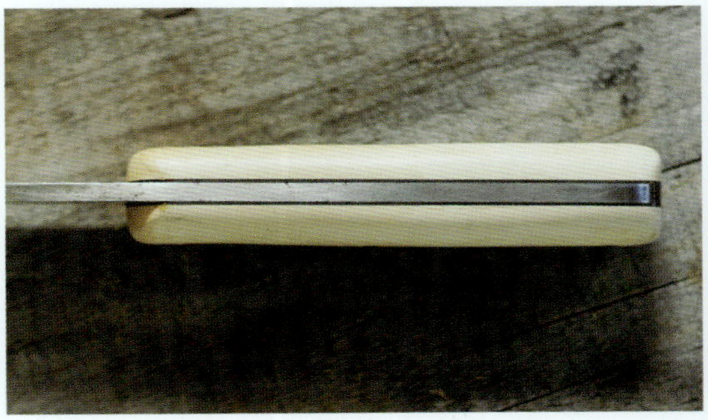

Spacers, also called liners, are thin pieces of material that are sandwiched in between the tang and the handle. This optional addition can be useful if you have thin scales and need to add some bulk to your handle. They can also act as a buffer to ensure a good fit for your scales and can compensate for some of the shrink and swell of the material if it is affected by heat or humidity. These spacers are on a bushcraft knife made by Abe Elias.

The linen or canvas base can give a Micarta handle character. This knife made by Jeff Bailey has a Micarta handle.

Pins

Pins are the pieces of thin, round metal that are inserted through holes to help hold the scales to a full-tang blade. These pins, once finished, will leave a small circle of metal visible on the handle. Pins can be made out of almost any kind of metal, depending on your particular taste. A well-chosen pin can complement and add to the overall appeal of your handle.

Stainless steel, nickel silver, and brass are all common choices for pins. You can use welding rod, nails, pins from knife supply companies, or any other round stock. I love using copper for my pins. Copper, much like steel, goes through an oxidation reaction. Unlike rust, this actually creates a protective layer that prevents further corrosion. It does cause a green patina, but with frequent use it won't form. Copper is a soft metal, but I've never had an issue with it being too soft to do its job. While it's not the most popular choice, I love its earthy feel.

If the holes you've drilled are the same diameter as your pins, you'll need to thin your pins down. Take your pins to the belt sander and grind them down evenly using a 120 grit belt. To ensure material is taken off evenly from the circumference of the pin, insert it in a hand drill spinning in the reverse direction of the belt. Lightly pass the spinning pin over the belt. The drill acts like a lathe, and your pin will come out round. Test them in the holes until they have a good fit. They should be snug, but you don't want to have to use to much force to get them to fit. This could cause the scales to crack or break.

Some knifemakers will pein the end of their pin to further secure the scales. By applying gentle taps with a hammer, the metal on the end of the pin is flattened, mushrooming out just slightly. By having the end larger than the rest of the pin, it really holds the scale to the tang. This technique was a necessity before the development of strong epoxy, but it is largely done as a show of expert craftsmanship today. If you do choose to pein your pins, make sure to countersink the holes slightly on the outside of the handle. This way, the metal will fill the countersink and won't stick out above the surface of the handle.

A technique to create hidden pins can be used if you don't want them to be seen as part of your knife handle. By stopping the drill holes before they exit the outside of the wood and cutting the pins short, the pins will still perform their job without being visible.

Thin your pins on a belt sander, making sure to keep their round shape by using a drill. If you decide to make more than one knife, you can invest in the proper size clearance bit to avoid this step. These bits come in a variety of numbers and will create a hole big enough to accommodate a specific size pin. If your hardware store doesn't have number size clearance drill bits, you can order them from an industrial supplier.

Using epoxy along with your pins is essential for more than just creating a secure bond of the tang to the scales. It also acts as a waterproof seal to prevent corrosion from happening under your handle. Don't use a quick-set epoxy, as they tend to not hold as well and you'll need the extra time to set up your handle properly. Pigment can be mixed in with the epoxy to change its color.

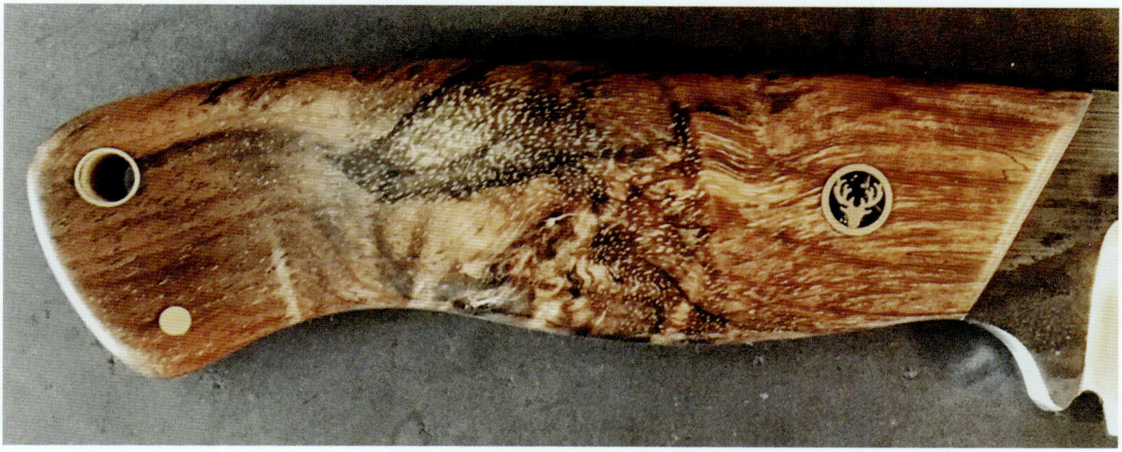

If you'd like to further customize the look of your blade using pins, mosaic pins can be used. Characterized by a unique pattern on the ends, these pins are fully functional as well as decorative. You can make your own by epoxying different sizes and shapes of thin diameter steel stock into metal tubing, or you can buy them from a supplier. These mosaic pins are utilized by knifemaker Mike Jones in his hunting knives.

Handle Ergonomics

A well-fitting handle should allow to you completely forget about the fact that you're gripping your blade at all. The size and shape of your hand will dictate what makes the best handle for you. More than being comfortable, good handle ergonomics will make you more efficient. If you're strained and using effort when you're holding your knife, you'll be more prone to fatigue, especially if you're using a knife for long hours or heavy work.

When knives are designed for production, the handle is crafted to fit the average hand of the consumer. The average hand of an adult male is about 4 inches (10.2 cm) across. As a smaller female, with a hand that measures 3 inches (7.6 cm) across, this creates a problem. While I can deal with a handle that extends well beyond my palm, the girth of the handle creates a bigger problem. While the average man tends to be in the 7½ inch (19.1 cm) range in length, my hand is a full inch shorter. My fingers usually can't wrap fully around the handle of most production knives, causing me to have to work hard to keep my hand on it. You should be able to easily close your fist around the handle without having to use a death grip.

I have a good friend who has the opposite problem—his hands are so massive, they make most knife handles disappear. Your handle should be long enough to allow you to use your entire hand to grip the blade, and the scales of your handle should curve out enough to fill up the space made inside your palm when you make a fist. The hand is a curved surface, so avoid straight lines and rough corners.

Try going to a store and holding a bunch of knives to see what shapes you like, and then work with that shape and modify it to fit your specific hand. While finger grips seem like a good idea, they can actually spread your fingers out and make for an awkward grip on the blade. A curve on the bottom of your handle will allow your fingers to wrap around nicely, and a tapered end can help maximize your grip. Your pinkie finger creates a much smaller grip circumference than your longer fingers, so this taper allows for the use of your full hand. Additionally, having a handle that is taller than it is wide will prevent the blade from spinning in your hand.

Think of the grip that you'll most frequently be using when you use your blade. You use a kitchen knife very differently than one that is designed to carve wood. Make sure your handle feels comfortable using this grip. Avoid making your grip too specific, however, as you'll hold it differently when completing different tasks.

Your handle ergonomics are correct if your knife feels comfortable in your hand. Take the time to really make sure you have a good fit; it's worth it in the long run. As the point of contact between me and my blade, the perfect handle makes me feel like my knife is an extension of my body and helps me to use my tool in a safe and effective way.

The minimalist design of this Bird and Trout knife by Patrick Durkin Cummins, appropriately named the Durkin Trout, makes for a secure grip without the addition of handle scales. This creative approach for a handle is great for knives that perform a specific task but can be limiting when utilized on a more multi-purpose knife.

Making a Wooden Handle

Despite all the different composite materials available, I love the comfortable feel and traditional look of a wooden handle. As you set up your work space to make your handle, be aware that the epoxy might make a mess. It might be a good idea to put down cardboard, newspaper, or butcher paper to protect any surfaces. Also, find a disposable container to mix the epoxy in. You can use an old plastic container or even a couple of pieces of painter's tape stuck to a piece of cardboard. Make sure your pins are sanded down to size before you start.

MATERIALS AND EQUIPMENT

- wooden handle scales
- pins
- 2-part epoxy
- permanent marker
- acetone
- painter's tape
- drill bit (to match pin size)
- C-clamps
- rubber gloves
- craft sticks
- hack saw
- wood saw or band saw
- belt sander
- sandpaper

Mark the scales to show how they will sit on the tang in a way that is clear to you. Decide on a system that works and use it for every knife you make.

1. Mark your scales.

Check to make sure your scales are perfectly flat. Figure out how you want your scales to sit on the blade, making use of whatever part of the grain you'd like to show. Make a mark on your scales to designate which side will be the inside and what will be the top and bottom. This will make it easier to keep track of how to put it together later on.

2. Drill your holes.

Position one of your scales as it will sit on the blade. Tape the blade and the scale together with painter's tape. Holding them secure, drill through the hole in your tang all the way through the scale. Push one of your pins through the holes, securing the two pieces. This will keep the existing holes lined up as you drill your second hole. Drill through the second hole in your tang. If you have more than two pins in your handle, repeat this process, securing each new hole with a pin. Don't forget to drill out your lanyard hole, if you have one. Using your marker, trace the outline of your tang on the inside of the scale.

3. Repeat the process on your second scale.

Remove the pins, the tape, and the first scale from your blade and put it aside. Tape your second scale to the blade and repeat the process.

4. Cut out the handle scales.

Heading to your band saw or wood saw, follow the lines you made to cut out the rough shape of your handle. Put the scales back on the knife, and sand down the profile of the handle to match the profile of the tang. The scales will be covering more of the bolster than you want at this point, so use your marker to draw a line where your handle will end.

5. Finish the bolster area of your handle.

Take the scales off of your blade and pin the two scales together. Cut off any excess material around the front of the bolster and sand it down to your designated line. Unlike the rest of the handle, this bolster area will be harder to work once the scales are attached, as the finished blade will protrude from the scales here. Working on it later will most likely result in scratching the ricasso, so sand the bolster area of your scales down to a fine finish grit.

6. Prepare the handle materials and mix the epoxy.

Using coarse grit paper, rough up the inside of your scales as well as the outside of your tang. Then, use acetone and a rag to wipe down all the surfaces that will be adhered. Epoxy won't bond properly to a surface if it is dirty or has any oils on it, so make sure your hands are clean as you're handling it. Following the instructions on your particular brand of epoxy, mix it together. Using a craft stick, spread it all over the inside of each scale. You want to make sure you have enough epoxy that when you clamp the scales together, all of the surface area on the inside is completely covered.

Make sure the handle scales and tang don't slide or move as you're drilling your holes, as everything needs to line up to ensure a good fit for the pins.

Bring the area on the front of the bolster to a finish before you attach the scales. This area will be almost impossible to work after your handle is assembled without scratching the polish on your blade.

(continued)

7. Assemble and glue the handle.

Put the pins in the first scale and fit the blade over the pins. Fit the second scale on, gently tapping the pins in to make sure they're in the right position. Using your C-clamps, clamp down on the handle, squeezing everything together securely. Wipe off any excess epoxy that runs out, but don't worry about getting all of it, as you can clean it up later. Let your knife sit clamped overnight to dry.

8. Sand the handle.

Once your epoxy has set, bring your knife back to the grinder. Clean up the edges, and keep shaping your handle. Test your grip as you go, taking off more material wherever it puts pressure on your hand in an uncomfortable way. Once you're happy with your grip, take the belts down to a finer grit. Polish your handle by hand sanding with finer and finer paper until you get to finish you want.

9. Finish your handle.

Finish your handle by applying a coat of oil to the wood. I prefer to use tung oil, but any wood finishing oil will work. Oiling your handle not only gives it a nice finish but protects the wood from moisture.

If in doubt about whether or not your epoxy has set, let it sit for longer. It's better to be safe than sorry at this point in the game.

You'll need the blade to be wrapped once again to make the sheath, but remove the painter's tape and take a moment to admire your first knife.

PRO TIPS: PAUL BRACH

LOCATION: CUMMING, GEORGIA

Paul Brach is a custom knifemaker who has been making knives for over thirty years. Paul enjoys the challenge of trying out different techniques and styles, and has made different knife styles from around the world. His talent and skill have earned him the level of a Journeyman Smith in the American Bladesmith Society. Paul shares his love of the craft with others by teaching knifemaking and giving demonstrations.

Knife style: "Sheffield-style Bowies or simple, clean hunting knives."

How he got his start: "As a young teen, I made a black powder rifle from a kit. I then put a handle on a pre-ground knife blade and was hooked. In about 1984, I began grinding my own blades. I recently added forging to my skill set."

Best knifemaking tip: "Avoid extreme shapes such as duplex grinds, saw teeth, or unusual grip angles or finger holes. Master making clean, functional knives first and then get creative with your lines as you begin to understand what works and what you are capable of."

On handles: "Make handles with a palm swell to aid in a secure grip. Avoid finger grooves or anything that will prevent multiple grip positions. In use, a knife has to work from different angles, including inverted while cutting edge up. Think through how your hand will engage with the handle through different cutting scenarios. The handle should also index well so that you can be certain where the edge is facing without looking at it. An oval cross section is a great way to make this happen."

A knife made by Paul Brach

CHAPTER EIGHT

Making a Sheath

No fixed blade is complete without a proper sheath. A sheath will prevent your knife from accidentally cutting or stabbing you or another object when it's not in use. It will also protect the edge from damage, and cushion the knife from any impact that could occur from a fall.

A sheath will allow you to safely and conveniently bring your knife with you wherever you go. The design of your knife sheath will determine how it is carried and how easily it is accessed. Consider where you'll be taking your knife and what you'll most likely be wearing to help you find the best style of sheath for you.

A good sheath design will allow you to safely and conveniently take your blade wherever you go.

Sheath Styles

A belt knife is designed with a sheath that has a clip or loop to allow you to wear the knife on your belt. A vertical carry is the most commonly used style of belt sheath and uses a loop on the top of the sheath to hold your knife vertically on your side.

This kind of sheath can also be made horizontally, with the sheath running along the length of the belt. The knife can then be worn anywhere along the circumference of your belt, on the front, side, or back. This style tucks the sheath away, preventing it from getting caught on brush or bumping into things. It also makes your knife less visible than a vertical carry. If you use a horizontal carry, make sure you have a way of securely holding the knife inside the sheath.

You can also make a canted sheath. This style involves attaching your belt loop at an angle so that your knife is carried on your side at an angle halfway between a horizontal and vertical carry. This can help keep your knife out of the way. If you make this style of sheath, think about what side you want to carry your knife on to determine which way the sheath will tilt.

If you don't wear a belt, consider using a clip on your sheath so you can wear it inside your waistband. This can also be useful to make your knife a concealed carry.

Neck knives are another popular way to carry a blade. A sheath for a neck knife is often made of a synthetic material such as Kydex, which offers a secure fit. The knife can then be suspended from a cord around your neck with the handle pointing down. This allows easy and quick access. A neck sheath can also be made from leather and worn with the handle facing up.

If you wear a canted sheath on your nondominant side, you can grab your knife with your dominant hand by reaching across your body. This is called a cross draw and can make for a comfortable and easy way to access your knife.

Attaching a clip to a sheath will allow you to use either your belt or a waistband to carry your knife.

Boot knives aren't as common as some other methods of carry but still provide an interesting way to carry a blade. Boot knives are usually small self-defense knives that are clipped or strapped into a boot. I used to travel with a small concealed boot knife back in my hitchhiking days, and it always made me feel really tough. It also tended to rub against my ankle and was never comfortable or practical to draw. A boot knife can make a great backup knife and is sure to spark a conversation if someone happens to see it.

You can get creative with designing a sheath to accommodate your needs. I've seen shoulder holster knives designed to hold a knife under a jacket and sheaths made to strap to the outside of a backpack. Dive knives, used by scuba divers and free divers, are designed to clip to an arm or a leg and hold the knife secure during underwater activities. I find a lot of women's clothing to be less than practical when it comes to carrying knives and have had to use creative solutions to wear my blade. I've even worn small knives clipped to the inside of my bra so that I didn't have to leave my knife at home.

A neck knife can provide a convenient alternative to a belt knife.

Sheath Materials

Pick a material that will be functional for the kind of sheath you want to make, and don't forget to make something that complements your knife and is easy on the eyes. These are some of the more common materials used in making knife sheaths.

Leather: Leather is a traditional and classic sheath material. It makes a sturdy sheath that is both beautiful and durable. The moldability of leather allows it to create a custom fit to your knife, and its softness makes for a quiet draw. Depending on its environment, leather can have a tendency to dry-rot, crack, or even mold over time. Regular oiling and care can prevent these factors from being a problem.

Kydex: Kydex is a synthetic material often used for knife sheaths. It is incredibly tough and durable and offers fantastic protection to your blade. As a thermoplastic, it can be molded exactly to the shape of your knife and provides a strong, tight fit. It isn't as quiet as leather on the draw, and the tendency of the blade edge to rub across the material can dull your knife over time. Kydex will stand up well to harsh environments or saltwater and will maintain its integrity without maintenance. It is available in a variety of colors and patterns.

Nylon: Nylon makes a soft, lightweight sheath. Nylon isn't as rigid as other materials, which makes it comfortable to wear but also means it won't offer as much protection for your blade. Nylon will stretch out over time and won't offer the custom fit that you'll get from leather or Kydex. Velcro or snaps are usually used to keep the blade from falling out. Nylon is an inexpensive material, but it will eventually wear out and can be vulnerable to slices or tears.

Plastic: Plastic is often used as a sheath material on cheap production knives. It is lightweight and inexpensive but tends not to make for a great fit. I have never had much success in keeping a knife in a plastic sheath, but it does offer good protection for your blade and is better than nothing.

Taking care of my knife when I'm using it on a survival expedition is a top priority. I often use the sheath of my knife to hang it in a specific spot in camp, where I know it is safe and I can locate it even in the dark. I was once on a survival trip in the deciduous forest of New England during autumn. I lost the dark, leather sheath for my knife in the blanket of colorful leaves on the forest floor. Knowing how important it was to maintain the safety of myself and my blade by having a sheath, I improvised and made one out of birch bark. The white bark had the added bonus of being seen more easily at night.

Proper maintenance of a leather sheath can help it last a lifetime.

A Kydex sheath with a horizontal carry design. Made by Jeff Santos.

The use of a scratch compass (left) and spacer set (right) can help you easily create uniform lines and holes in leather. Using these tools can help you create a better-quality leather sheath.

Leatherworking Tools

While it's possible to improvise, buying a couple of inexpensive leatherworking tools is a great investment if you plan on making leather sheaths. The time and frustration I save by using these tools make them worth their weight in gold.

Scratch compass: This compass is set at a specific measurement and dragged along the edge of the leather, marking a line with consistent spacing. I use the scratch compass for marking the stitching line of my knife sheath.

Spacer set: This handy little tool is used to mark consistently spaced stitch holes. Yes, you can absolutely measure and mark out each hole on the leather of your knife sheath. Or, you could roll the spacer set along the line you made with your scratch compass and instantly have marks for perfectly spaced stitch holes.

Making a Leather Sheath

The natural feel and durability of leather makes it one of my favorite sheath materials. The following instructions will give you the process of making a simple and effective leather sheath.

MATERIALS AND EQUIPMENT

- paper
- tape
- plastic cling wrap
- 7 oz to 8 oz (207 ml to 237 ml) vegetable-tanned leather
- artificial sinew
- leather dressing
- contact cement
- pencil
- scissors
- ruler
- utility knife
- matches
- pliers
- drill press
- 1⁄16" (2 mm) drill bit
- spacer set
- scratch compass
- harness sewing needles
- wood file
- belt sander
- sanding belts (60, 120, 200 grit)
- leather dressing

1. Draw the pattern.

Using your ruler, draw a straight line the full length of the paper. This will be the leather seam of the sheath. Place the spine of your knife about ¼ inch (6 mm) from the line and roughly parallel with it. Then, trace around the curve of the knife edge, leaving a bit of space between the actual edge of the blade and your pencil line. Extend this line up two-thirds the length of the handle and make a mark to designate the top of your sheath. Draw a second line running along the knife edge, but make this one about ½ inch (1.3 cm) out from the first line. The space in between these two lines is where your welt will go. At the top of the pattern, extend the two lines at least 5 inches (12.7 cm) past the top of your sheath to create the belt loop. I like the belt loop to be more narrow than the sheath, so taper both sides so that the loop will not be visible when you view your sheath from the front. Make sure you leave enough length so that when you fold and stitch the strap, you have a space that is at least 2 inches (5.1 cm) wide for your belt. Fold the paper in half on the center seam and cut around your pattern through both sides of the paper to give you two identical sides of your sheath. Trim the sheath so that only one side has a belt loop.

2. Test the pattern and draw the welt.

Place your knife in the sheath pattern and see if there are any modifications you need to make. Leather is thicker and more cumbersome than paper, so depending on how thick your leather is,

(continued)

The exact dimensions of your sheath will be relative to the size of your specific blade.

Be careful not to overshoot the cuts on the corners, as this will cause a weakness in the sheath.

you might need to make your design bigger if the knife isn't fitting well in the paper. If everything looks good, trace the curve of the sheath on a leftover piece of paper. Then, trace the inside line to mark out the pattern for the welt. The welt is a strip of leather sewn between the two sides of the sheath, creating a boundary between the knife's edge and the stitching. Cut this pattern out and test it out in your paper sheath. The welt should fit neatly inside the seam of the sheath.

3. Transfer the pattern to the leather.

Place the pattern on the rough side of the leather. This will be the inside of your sheath. It's important at this point to think about whether you're making a right or left handed sheath. If you're making a right handed sheath, you should have the belt loop side of the pattern on the left side of the sheath when it is placed on the leather. If you're making a left handed sheath, flip the pattern over to put the belt loop on the right side. Tape the pattern down and trace it with a pencil. Cut through your lines using a sharp utility knife. When cutting out the welt, leave a little extra leather on the outside edge. It can be sanded off the finished sheath, and this way you can be sure the leather ends up flush. When you have everything cut out, place your knife as it will sit in the sheath. Try wrapping the leather around it to make sure everything fits together and that there will be enough room for your knife.

4. Make belt stitch holes.

Fold the belt loop down to where it will be attached to the sheath and trace around the piece of leather that will be stitched down. Using your scratch compass, make a mark ¼ inch (6 mm) in from the edge of the leather at the end of the belt loop where the stitches will be going. Roll the stitch set around the seam, following the line you made with the compass. You'll need the holes to outline the entire area that will be in contact with the sheath. Use the drill press with a ¹⁄₁₆ inch (2 mm) bit to drill through the marks on the leather.

5. Glue the belt loop and redrill the holes.

By using glue in addition to stitching, you can make sure all the seams of your sheath are tough enough to stand up to whatever you put them through. As a bonus, it makes it way easier to keep everything in place later on as you drill and stitch. Use your file to rough up the smooth part of the leather where the belt loop will be attached to the sheath, as well as the side of the loop that will be stitched down. Apply contact cement on both pieces of leather and let it dry for a bit according to the instructions on the bottle. Press and hold the two pieces together to make sure they bond. Redrill the holes you made in the belt loop, this time going through both pieces of leather.

6. Stitch the belt loop.

Use a basic running stitch to go through all the holes, and then stitch back through the alternating holes to fill in the spaces. It can be tough to work the needle through the leather with your fingers, so I like to grab the needle with a pair of pliers and push and pull it through that way. It's easy to break needles doing this, so keep several spare needles on hand. Every two or three holes, tighten your stitch by pulling on the thread. To end the thread, go back through a few stitches and then tie a knot. Melt the end by burning the thread with a match.

7. Glue the seam and the welt.

Once the welt is glued in, it will be almost impossible to remove it to make adjustments—so test your sheath with the welt lain inside it before you commit with the contact cement. Make sure your knife has enough space to lie inside the sheath. Mark where the edge of the welt will lie by tracing it with a pencil. Use your file to rough up the smooth side of the welt. Apply glue to the area you traced where the welt will be lying, as well as the same area on the opposing side of the sheath and both sides of the welt. Wait for the contact cement to set, and then press all three layers of leather firmly together. At this point, it's really starting to look like something!

(continued)

Use a file to rough up the leather. This creates a better surface for the contact cement to hold on to.

The running stitch is a very basic stitch that most people are familiar with. The needle is brought back and forth through the material, alternating sides with each stitch.

8. Sand the edge.

Use the belt sander to sand down the edge of your sheath so that all three layers of leather are flush. Take it down to a fine grit and soften the sharp edges of your grind a bit, which will pretty things up. As you define the edge of your sheath, you'll have a definite edge to work off of when marking out your stitch holes.

9. Make the stitch holes in the main seam.

Use the scratch compass to make a line around the entire seam of the sheath about ¼ inch (6 mm) from the edge of the leather. Use your spacer set to mark out the holes for drilling and drill through all three pieces of leather to make the stitch holes.

10. Stitch the sheath.

Cut your thread four to five times the length of your sheath. Starting three holes from the top of the sheath, thread your needle and thread through the hole. Pull the end of the string and even the two sides up so that half of the string is on one side and half is on the other. Thread a needle on the other end of the thread so that you now have two needles to work with. Stitch back up with one needle through the holes you missed using a running stitch. When you reach the top, come back down the sheath until you get back to the place that you started. Do the same thing with the needle on the other side, going through the same holes again. This creates a backstitch that reinforces the upper corner, which is the part of the sheath that takes the most abuse and is the most vulnerable to wear and tear. Once you're back where you started with both needles, start stitching down the sheath. Go for two or three holes before switching to the other needle and filling in the spaces.

11. Add the final touches.

Soak the leather in warm water for five to ten minutes or until it starts to soften. Wrap the entire knife in cling wrap and secure the wrap with tape. This will keep the water off your blade and prevent it from rusting, as well as give it a little more bulk so that it will come out of the sheath easily even after the leather has dried and shrunk. Place the knife in the leather and let it dry around your knife. Once your sheath is completely dry, remove your knife and take it out of the plastic wrap. Try removing and inserting your knife a few times to see if there's a good fit. If you're not happy, rewet the sheath and try forming it again. Leather dressing adds water resistance to the leather and makes rewetting it to reform it much more difficult, so make sure you like the way your knife fits before you go to the next step. Once you're happy with your sheath, cover the outside with a few coats of leather dressing and admire the work you've done.

Fill in the spaces that don't have thread by using your second needle. Don't forget to give your stitches a tug every few holes to keep things tight.

The finished leather sheath for the project knife

CHAPTER NINE

Sharpening and Knife Care

Sharpening the edge of your blade is the last step in making your knife. Like most things in knifemaking, every maker has their preferred method for sharpening. In my opinion, the best method for sharpening is one that leaves you with a useful knife. A knife is only as good as how well it performs its function, and there are very few knives that were designed to be dull.

After all the work that you've put into making your knife, it's almost time to head out into the world and put it to the test. With proper care, a knife can last a lifetime. Learning to safely and effectively use your blade can allow you to get the most out of your functional work of art.

Hand-sharpening a blade in front of the shop. Sharpening is almost a form of meditation, requiring patience and controlled consistency.

Sharpening Tools

The sheer number of different sharpeners available on the market can be overwhelming. In my opinion, the confusion this can create is part of the reason so many people carry dull knives. Different sharpeners require different techniques, but the principles of sharpening remain the same. The following are some of the more common styles of sharpeners available.

Just a handful of the many sharpening tools available on the market

Stones

There is a variety of different types of stones available for sharpening, and each has advantages and disadvantages. Some stones require oil or water to be used as a lubricant. This lubricant reduces the friction of grinding and helps to remove the swarf to keep the stone working effectively. *Swarf* is the name given to the small metal dust that is abraded off your blade during the sharpening process.

Each stone will be a specific grit. You'll need to be able to move from a coarser stone to a finer stone to complete the process, so get a few stones of varying grits.

A stone is the tool most people picture when they think of sharpening a knife.

Waterstones

Waterstones require water to be used as a lubricant. They are the softest of the sharpening stones, and as they are used, the surface is constantly wearing away. This exposes new, fresh abrasive and makes them great for sharpening. This stone dust, or slurry, also works to polish the bevel. The constant wear creates an indentation in the top of the stone, which needs to be flattened frequently to maintain a good surface for grinding.

Japanese waterstones: Japanese waterstones are a kind of natural waterstone that is mined in Japan, and they need to be soaked in water for twenty-four hours before they can be used. They are widely

Stones that are made for use on a flat surface are called bench stones. Smaller stones that are handheld are called pocket stones. This diamond pocket stone is carried by the author in the field and has stood up well to the abuse of traveling around the world.

renowned as some of the best-quality stones in the world but are rare and, as a result, expensive.

Synthetic waterstones: Synthetic waterstones are made by combining aluminum oxide with resin. While some die-hard waterstone users prefer the natural stone, most professionals agree that synthetic waterstones perform just as well.

Oilstones

Oilstones are harder than waterstones and require oil instead of water as a lubricant. This can be a bit messier, and it can be less convenient to have oil readily available to use, especially in the field. They don't wear down as fast as waterstones, but they also don't polish quite as well.

Arkansas stones: This is the most common kind of natural stone and is a type of novaculite that is mined in the United States. It comes in several grades, from coarse to fine. They tend to cut a bit more slowly than man-made stones but can produce a nice polished edge.

Aluminum oxide oilstones: Also known as India stones, these synthetic stones are made from aluminum oxide and resin. Unlike synthetic waterstones, the resin used is much harder and doesn't wear as quickly. The quality of these stones can vary greatly, but they are less expensive than Arkansas stones.

The Work Sharp sharpener is an example of an electric sharpener designed by world-renowned knifemaker Ken Onion. It has a variety of different belts for grit progression and puts an excellent edge on a blade.

Silicon oxide oilstones: These stones are the fastest cutting of the three kinds of oilstones. They don't make quite as fine of an edge but are great for extremely dull knives or for starting your edge before progressing to something finer.

Ceramic Stones

Ceramic stones are made by combining ceramic powder and aluminum oxide. They don't require any lubrication to use and tend to be more expensive than a lot of stones, but don't wear quickly and can last a long time. They can be brittle and usually are only available in finer grits.

Diamond Stones

These stones utilize small, man-made diamonds to sharpen the steel. They don't require any lubrication, and because the surface isn't being abraded, they don't lose their flatness. While convenient and fast, they don't create quite as good a polish as waterstones.

Guides

There are a number of different guides available on the market designed to hold your blade at a specific angle. You can then run your blade across a stone and avoid the learning curve that comes with figuring out how to maintain a consistent angle.

Electric Sharpeners

Electric sharpeners encompass a wide variety of devices designed to sharpen a blade. They can range from high-quality, effective sharpeners to something that will destroy the edge on your blade. You often get what you pay for, so do your research before buying one.

V Sharpener

These sharpening tools have two edges, shaped in a V. A blade can be moved through the V with light pressure to grind the edge. By holding the sharpeners at a specific angle, it creates consistency in every pass that can be hard to obtain using a stone. Unfortunately, it most likely isn't the exact angle you want. It also won't compensate for the loss of steel that occurs with use over time and will change the shape of your edge.

Crock Stick Sharpener

These sharpeners have a fixed base, with abrasive ceramic rods sticking out that are fixed at an angle. The knife is sharpened by running your knife against the rod. Although they are designed to hold a consistent angle, they can be difficult to use, depending on the specific grind on your knife. I find they can work to touch up an edge but don't work well to redo an edge on a knife that is very dull.

Belt Grinder

The same grinder that you used to make your grinds can be used to sharpen your edge. The knife is held gently at the appropriate angle against the slack part of a fine belt. This takes a good eye and a steady hand and can be difficult for a beginner due to the speed of which the metal is removed. Grinding in this manner tends to create a slightly convex grind.

Buffer

A buffer is a machine with a soft, rotating wheel that can be used on both your handle and your blade. By using different wheels and compounds, you can control the level of abrasiveness to get your desired finish. Using this to finish your edge can create fantastic results. It's important to use extreme care if you decide to buff your knife. I use a large shop buffer when I'm making jewelry and have learned firsthand why the buffer is beyond a doubt *the most dangerous tool in the shop*. The slightest edge can catch, grabbing and flinging what you're grinding with incredible speed and force. It's bad enough to shoot a small dog tag across the room, but that's nothing when you think about the potential consequences of rocketing a wickedly sharpened knife. If you choose to use a buffer, practice with something that is less likely to kill you first.

Mechanics of Sharpening

The thinness of an edge makes it the most vulnerable part of your knife. This is also the part of the blade that takes the most beating. Every knife will require edge maintenance eventually, as even the best steel will wear with time. The basic mechanics of sharpening will remain the same, whether it's your blade's first edge or its hundredth.

Some grinds do best with a small, secondary bevel on the very edge. Other grinds, such as the Scandi grind, are sharpened by refining the original grind. This makes the Scandi grind a very easy grind to sharpen for a beginner, as the angle needed is easy to determine. Depending on the design of your blade and the thickness of your steel, the exact angles you are creating will differ. The edge geometry you used in selecting and creating your grind apply for sharpening as well—just on a finer scale. The first step in grinding is to have a clear picture of the results you are looking for.

Make sure you have good lighting before you start the sharpening process. As with grinding, your sharpening process involves starting with a coarse grit and slowly moving down to finer and finer grits. You'll have to be able to see when you've removed your lines from the previous grit, which can be difficult when you're working on such a small scale.

If you are putting a new edge on your blade after use, you'll have to scrape away the worn, damaged old steel on the edge first. This will expose the fresh, strong new steel underneath. You'll then make a new edge using this steel. Inspect the blade and see what you're working with. Use a magnifying glass if you are having a hard time seeing the edge. Look for any small chips and assess any visible damage.

When sharpening, the key is to match the angle of the knife's edge to the sharpener. By keeping this angle consistent and moving your edge across finer and finer grits, you'll remove all the metal that won't make up the edge of your blade. The mechanics behind sharpening aren't difficult to understand, but good results take a skilled hand and a fine attention to detail.

If your blade has small chips, you'll have to grind the edge all the way down with a coarse grit until they're gone. Returning to the belt grinder might be the best option for severe chips or broken blade tips.

When a knife is dull, you rely less on the edge geometry for cutting power and need to use more physical pressure to get the job done. This increases your likelihood of slipping and accidentally cutting yourself. Combining that with the added force you're using, and you can really ruin your day.

The scars on my left hand were a result of very good knives with very poor edge care. I learned my lesson the hard way, and they now serve as a reminder as to why it's so important to maintain the cutting edge of any knife.

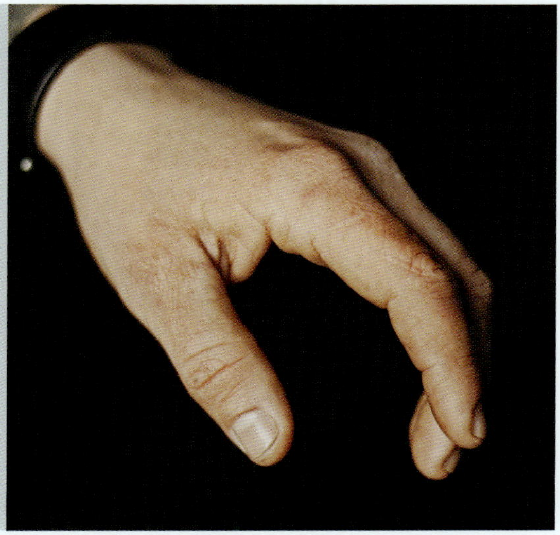

The most important part of sharpening is consistency. If every sharpening stroke you take is at a different angle, you can actually dull your knife. A sharp blade has clean, symmetrical angles. Freehand sharpening takes practice and patience. Use slow, steady hands and take your time. Choose a method and routine for sharpening and stick to it. Repeating your process over and over will commit it to muscle memory and make it second nature.

The level of grit you use to finish your edge is entirely up to you. Grinding creates a scratch pattern, and when sharpening, your edge will have tiny micro-serrations. The coarser the grit of your final grind, the bigger these micro-serrations will be. While the size of these serrations becomes important if you are making high-end specialty knives, the average user won't notice much, if any, difference on this level.

Sharpening a Blade

While using a belt grinder to sharpen your blade is a great method, it also requires actually having the grinder available anytime you need to put a new edge on your blade. As someone who travels frequently and is often away from electricity, I don't always have this luxury. Learning to use a stone is a worthwhile time investment that will give you the freedom to maintain your blade anywhere.

One useful technique to use when you're learning to sharpen is to use a marker to color the edge of the blade. Much like marking the steel during the grinding process, this will allow you to see where you are removing metal and where the metal isn't coming in contact with the sharpener. It will also allow you to see the angle at which you're grinding.

A strop will be used to make the final touches on your edge. Stropping will remove the last of the burr and clean and align your blade's micro-serrations. You can buy a strop specifically made for this purpose, or you can use an old belt, a piece of denim, or even cardboard in a pinch.

MATERIALS AND EQUIPMENT

knife

sharpening stone

strop

Use a marker to help perfect the angle that you're holding the blade at in relation to the stone and check to see where you have removed your marks. An angle that is too steep will only get rid of the marker on the very edge where the bevels meet. An angle that is too flat will remove your marker from the side of your edge, leaving the marker closest to the edge.

1. Grind the first bevel of your edge.

Hold your knife flat on its side on your coarse stone. Lift the spine up slightly so that the edge is resting on the stone at a sharp angle. Move the edge across the stone lightly, as if you were trying to slice off a very thin piece of it. Make sure the entire length of the blade, from the heel to the tip, comes in contact with the stone. Repeat this process several times, maintaining the same angle.

2. Check for the burr.

As you remove steel from the edge, a tiny burr will eventually form on the opposite side of the edge. The burr is a rough, raised metal curl that results from grinding metal. The burr should appear evenly along the entire length of the edge If you find that it is absent in an area on your edge, you aren't grinding as much on that particular spot.

3. Switch sides and repeat the process with a finer grit.

Once you have a burr along the entire edge, switch sides. Repeat the process until you have a burr on the second side. If you have any chips in

Bring the edge of the blade across the sharpener lightly. Let the stone do the work instead of trying to muscle through it.

The burr might be visible, but you can also test for it by using your fingernail. Scratch the edge of your blade gently, from the spine to the edge. If your nail catches slightly, you have a burr.

your blade, you'll have to continue grinding with a coarse stone until all the steel is removed past that chip. Once you have set your edge with the coarse stone, move to a finer grit stone and repeat entire the process again.

4. Strop your edge.

Use your strop to remove the final burr on the blade's edge. Stropping involves the same motion that is used while sharpening on a stone, but in reverse. Instead of cutting forward, the blade is drawn back, dragging the edge on the strop. Use light pressure and make several passes, alternating sides.

5. Test your edge.

There are many ways to test your edge to see if it is sharp enough. Everyone has their own favorite method, from seeing how well the blade cuts through paper to carefully drawing the edge along a fingernail. In my opinion, the best way to see if an edge is sharp enough is to try using it for its intended purpose. A properly sharpened knife will complete the task it's designed for with ease, so see how well it performs in the field. As you become more familiar with your particular blade, you'll be able to tell when it starts to lose its edge.

(continued)

Stropping helps to align the blade's edge and can be used to maintain your blade between sharpenings.

Skinning bark from a tree with a bushcraft knife. Using the blade for its intended purpose and seeing how well it functions is a great way to determine if its edge is adequate.

In the survival world, I often see people trying to sharpen their knives on flat rocks while out in the field. I strongly advise against this practice unless it is absolutely necessary. While this technique has certainly been used for centuries, the quality of the steel in a well-made knife today will hold a decent edge even with tough use over an extended period of time. When sharpening, you rely on near-perfect flatness on an almost microscopic level to create a good edge. The slightest ridge or tiny bump can be enough to further dull your blade. I have traveled all around the world and have very, very rarely found a stone flat enough to improve the current edge on my blade. This technique should only be used in dire circumstances, as in many cases it can do more harm than good.

USING A HONING STEEL

Working as a butcher, I would use a honing steel many times throughout the course of a day to maintain the sharp edge on my knife. It is a common misconception that this stick-like steel rod, found in many kitchens, is used to sharpen a blade. Unlike sharpening, which removes material to reset an edge, honing is used to recenter the knife's edge. Through use, this edge can accumulate tiny bends and warps that affect the sharpness of a knife. Running the blade at an angle along steel pushes the edge back in line. This process is especially useful in maintaining the edge of a very thin and sharp blade, such as on a kitchen knife.

It requires a great deal of skill and practice to use a honing steel in the air as many chefs do. It is much easier to get a correct angle by resting the end of the steel on a flat surface and running each edge of your blade along the side of the steel at an angle.

Keeping an Edge

Knifemakers are at an advantage when it comes to knowing how to take care of a blade. Now that you know what it takes to make an edge, it's easier to imagine and avoid the processes that will ruin one. If you're in a survival situation, it's incredibly important to protect the item that you are using to keep you alive. Even if you're not, it can be devastating to ruin something you've invested so much time into.

My number one pet peeve with knives is people who carelessly stab their knife into the dirt when working outside. Even if they don't stab it on a rock and chip the edge, soil is an abrasive. The edge of a blade is created with abrasives; it can also be destroyed by abrasives. Keep your knife in its sheath when you're not using it. This protects your blade from any unnecessary damage, as well as protecting you from accidentally cutting yourself.

Avoid cutting on any cutting surface that is harder than your blade's edge. Another common practice I see is using a rock as a cutting surface. Your blade will thank you if you use a log or piece of wood instead. Cutting boards not only serve as a surface to cut on but are important to protect the integrity of the knife.

It's tempting to use your blade as an all-purpose tool, but take it from someone who learned it the hard way; your knife is not a pry bar or a screw-driver. You might get away with it for a while, but if you make it a habit, your blade will probably take damage at some point.

Make sure to dry your blade when it gets wet and rinse it if it comes into contact with saltwater. Don't store your blade for long periods of time in the sheath, as the leather can collect moisture and cause corrosion. All carbon steel will need care to prevent corrosion in the form of surface rust, so anything you can do to be proactive is a help. If you do see rust forming, use fine grit paper to remove it as soon as possible and apply a light coat of oil. I also recommend oiling your blade every so often, especially if it will be sitting unused for a period of time.

I let a good friend borrow my favorite blade in the jungles of Vietnam. He had never used a knife on giant bamboo before and had no reason to think my knife wouldn't be able to handle it. Giant bamboo is tough, thick, and surprisingly dense. When you chop into it, it can grab and hold on to the edge of your blade, particularly if you cut into a node. If you deviate in the slightest when pulling it out, it can chip the steel. Needless to say, my blade came back with an unintentionally serrated edge; and after seeing the heartbreak on my face when he apologized, I'm sure he will never look at bamboo the same way again.

PRO TIPS: MIKE JONES

LOCATION: QUESNEL, BRITISH COLUMBIA

Mike is a custom knifemaker based out of Canada, and he makes all his own knives from start to finish. In true Canadian fashion, he often uses reclaimed steel from old sawmill blades to make his knives, perfecting his own heat treatment through experimentation. He is known for designing a tool that is an innovative cross between a machete and an ax, and for his hunting knives that have proven themselves in the harsh wilderness of British Columbia.

Knifemaking style: "Functional tools that make you stoked to use them."

How he got his start: "I was at the night market in Vancouver. They had a table with two-for-one knives, and . . . my buddy [and I] both got two. They ended up being really poor-quality knives, and my friend showed me that one of his had a handle that had been screwed on. He said he was going to take off the handle and put a new one on. It never occurred to me that I could do that. I thought I could do the same with my knives, so I went on YouTube to find out how. I found all these videos on how to make an entire knife, and it seemed accessible, so I gave it a shot. I was blown away that you could just make a knife. It was so cool to me to think that any shape I could draw on paper I could make into a knife."

Best tip for a beginner: "Don't get hung up on what kind of equipment that you have in your shop. People make knives just with files, so don't let a lack of expensive equipment stop you from starting to make knives. You can get by with some pretty modest equipment."

On sharpening knives: "Understand how sharp your knife needs to be. People like to put a shaving edge on an ax, even if it's just going to be used for chopping wood. It's cool to shave your arm with an ax that has a mirror edge on it, but it doesn't need to be like that. Don't waste your time trying to get a knife hair-popping sharp when you're chopping up kindling with it. That's time that could be spent working on your next knife."

This unique blade made by Mike Jones is a cross between a machete and an ax and makes easy work of wood.

Glossary

Alloy: A metal formed by mixing two or more elements, at least one of which is metallic

Annealing: A form of heat treatment used to soften steel to make it more workable

Austenite: The phase of steel that occurs when it is heated at temperature above roughly 1,350°F (732°C)

Blade profile: The overall shape of the blade when viewed from the side

Bolster: The band between the handle and the cutting edge, which creates a transition from the handle to the blade and helps to prevent your hand from being cut on the knife edge

Burr: The rough, raised metal curl along the edge of a blade that results from grinding

Choil: The small cutout on the base of the edge of the blade closest to the handle

Critical temperature: The temperature at which the crystal structure in steel undergoes a phase change

Curie point: The temperature at which steel loses its magnetic properties

G10: A composite handle material made from fiberglass cloth and resin

Guard: The part of the knife between the handle and the cutting edge designed to prevent the fingers from slipping forward into the blade's edge

Hardening: A form of heat treatment used to harden the steel

Hardy: Tool with a square shank that fits into the hardy hole on top of the anvil

Hardy cut: A kind of hardy used for cutting through heated metal

Heat treatment: A broad term for a process used in the controlled heating or cooling of steel to cause specific changes in its properties, though often used in reference to hardening and tempering

Heel: The part of the blade that is closest to the bolster

Honing: A process used to maintain a knife by recentering the knife's edge

Jimping: The series of notches on the spine of the knife used to add texture for grip

Kydex: A synthetic sheath material

Martensite: The phase that steel is in when after going through the hardening process

Micarta: A synthetic handle material that is a composite of fabric in thermosetting plastic

Normalizing: Form of heat treatment used to create a uniform structure in steel

Pearlite: The phase of steel that results when it is slowly cooled

Platen: The plate behind the belt on a belt sander

Plunge line: The transition between the ricasso and the cutting edge where the bevel grind begins

Pommel: The cap on the end of the handle of some knives

Quench: The container for holding liquid to cool down your metal

Quenching: The process used to cool metal rapidly to cause it to harden, usually through submersion in oil or water

Quillion: Another term for the guard on a knife, often specifically in reference to the pieces projecting from the handle that stop the user's fingers from sliding onto the blade

Ricasso: The unsharpened length of blade between the handle and the start of the cutting edge

Scales: The two pieces of material that attach to the blade's tang to form the blade's handle

Scratch compass: A leatherworking tool used to mark a line around the edge of leather

Spacers: Thin pieces of material that are sandwiched between the tang and the handle of some knives

Spacer set: A leatherworking tool used to mark out consistently spaced stitch holes

Spine: The thick part of the blade that runs along its back and provides support for the knife

Stropping: The process of removing the last of the burr and aligning the micro-serrations on a blade's edge, used to finish a knife during sharpening

Swarf: The small chips of steel or stone created during the sharpening process

Tang: The unsharpened extension of the blade to which the handle is attached

Temper: A form of heat treatment used to take away the brittleness caused by hardening

Welt: The strip of leather sewn between the two sides of the sheath to create a boundary between the knife's edge and the stitching

Resources

Organizations

American Bladesmith Society
www.americanbladesmith.com

Australian Knifemakers Guild
www.akg.org.au

German Knifemakers Guild
www.deutsche-messermacher-gilde.de

Italian Knifemakers Guild
www.corporazioneitalianacoltellinai.com

The Knifemakers Guild
www.knifemakersguild.com

Schools

Bill Moran School of Bladesmithing
Texarkana CollegeTexarkana, Texas
www.texarkanacollege.edu

Carter Cutlery Bladesmithing Courses
Hillsboro, Oregon
www.CarterCutlery.com

La forge d'Ostiches School
Ostiches, Belgium
www.forge-ostiches.be/fr

New England School of Metalwork
Auburn, Maine
www.newenglandschoolofmetalwork.com

School of Blacksmithing and Bladesmithing
Welling, Kent, England
www.owenbush.co.uk

Tharwa Valley Forge
Tharwa, Australian Capital Territory, Australia
www.tharwavalleyforge.com

Magazines

***BLADE* Magazine**
www.blademag.com

Knife Magazine
www.knifeworld.com

Knives Illustrated
www.knivesillustrated.com

Suppliers

Alpha Knife Supply
www.alphaknifesupply.com

Blacksmiths Depot
www.blacksmithsdepot.com

BRISA
Finland
www.brisa.fi

Combat Abrasives
www.combatabrasives.com

Gameco Artisan Supplies
Australia and New Zealand
www.artisansupplies.com.au

High Temperature Tools and Refractory
www.hightemptools.com

Knife and Gun Finishing Supplies
www.knifeandgun.com

Lansky Sharpeners
www.lansky.com

New Jersey Steel Baron
www.newjerseysteelbaron.com

Pop's Knives and Supplies
www.popsknifesupplies.com

Red Label Abrasives
www.redlabelabrasives.com

Supergrit
www.supergrit.com

Texas Knifemaker's Supply
www.texasknife.com

Tru-Grit
www.trugrit.com

Vegas Forge
www.vegasforge.com

Work Sharp
www.worksharptools.com

Online Forums

Blade Forums
www.BladeForums.com

The Knife Network
www.knifenetwork.com

About the Author

Laura Zerra is a primitive survival expert who has been practicing her craft in different ecosystems around the world for the past fifteen years. She is perhaps best known for her appearance on Discovery Channel's *Naked and Afraid* and *Naked and Afraid XL*, completing survival challenges in Panama, the Peruvian Amazon, and Colombia. She spent a combined total of eighty-two days in the wilderness, relying on her knife as her sole survival tool.

Through her extensive time spent in these and countless other survival situations, Laura gained a deep appreciation for a good knife and an understanding of what elements truly work out in the field. She chose to master the art of knifemaking to not only create her own perfect knife, but also to experience surviving in the wild with a blade made by her own hand.

In addition to her survival experience, Laura has worked as a butcher, a taxidermist, and a farrier. Learning these crafts has further added to her knowledge, understanding, and love of knives.

Her belief that life is a continuous learning process drives her to constantly seek new experiences and push the edges of her comfort zone. A self-proclaimed nomad, Laura shares her experiences through speaking appearances and teaching. When not practicing the art of primitive survival or working in the shop, her hobbies include hunting, spearfishing, free-diving, bow building, shed antler hunting, and learning new skills.

You can follow her adventures by visiting www.laurazerra.com.

Index